The Unofficial Guide to Open Broadcaster Software

OBS: The World's Most Popular Live Streaming Software

By Paul William Richards

Copyright © Paul William Richards
All rights reserved. No part of this book may be reproduced in any manner without written permission except in the case of brief quotations included in critical articles and reviews. For information, please contact the author.

Photographs by Michael Luttermoser Cover Photography Copyright © Michael Luttermoser.

DISCLAIMER AND/OR LEGAL NOTICES Although the author and publisher have made every effort to ensure that the information in this book was correct at press time, the author and publisher do not assume and hereby disclaim any liability to any party for any loss, damage, or disruption caused by errors or omissions, whether such errors or omissions result from negligence, accident, or any other cause

ISBN: **9781098933845**

DEDICATION

To the contributors of Open Broadcaster Software and the online community in the OBS forums.

The Unofficial Guide to OBS

CONTENTS

	Acknowledgments	i
1	A community of live streamers	1
2	Video is the new frontier	6
3	The live streaming landscape	12
4	OBS live streaming course overview	21
5	Creative ways people are using OBS	93
6	Origins and development history of OBS	102
7	Popular topics in the OBS forum & top plugins	110
8	Audience engagement and community building	113
9	Conclusion	127

ACKNOWLEDGMENTS

To everyone who helped solve a technical issue or provided guidance inside our **OBS Users** Facebook group (facebook.com/groups/OBSUsers).

The Unofficial Guide to OBS

1 A COMMUNITY OF LIVE STREAMERS

Today more than ever Open Broadcaster Software is being recognized as a valuable video production tool in the broadcast industry. Amidst the halls of the 2019 National Association of Broadcaster's Show (NAB Show), where the world's highest end video production software and hardware is on display each year in Las Vegas, Nevada, you will find OBS is talked about quite frequently. After attending the NAB Show for the past six years, I have had countless conversations with OBS users from around the world. A common theme of conversation is always the "online community" and helpful forums. As a member of multiple online Facebook Groups including the Streaming Idiots group and OBS Users group, I can tell you the community is alive and well. Check out this picture from the 2019 NAB Show below during the Streaming Idiots Meet Up this year.

So what is OBS and why are all these people excited about it? OBS stands for Open Broadcaster Software, and it is the most popular free live-streaming software in the world. OBS is 100% open source which means the project is reviewed, maintained, and enhanced by a community of volunteers. Anyone can use it for free and participate in its development using Github, Dischord, or other online collaboration tools. If you find OBS valuable, consider supporting the development of OBS using Patreon or OpenCollective.

> **Pro Video Tip**
>
> Don't forget to check out the Glossary of Terms in the back of this book. Learning the video production vocabulary is the best way to get started with OBS!

This guide has been written to compliment an online OBS training course available on Udemy.com. With this guide, the included online course, downloadable materials, and a complementary audiobook, anyone interested in learning more about OBS should be able to advance their skills efficiently with these resources. As you start learning from these resources, consider learning from colleagues online in the OBS Facebook User Group available at: facebook.com/groups/OBSUsers.

The OBS suite is a versatile tool for live streaming and recording video. It can be used to record presentations, screen-capture sessions, eSports gaming, and much more. OBS includes a robust set of tools for processing audio which are discussed throughout this book. OBS can eliminate the need for expensive internal video capture cards with the integration of the NewTek® NDI®.

OBS was initially released in September 2012, by lead developer, Hugh "Jim" Bailey. The project has grown significantly over the years as new features and contributors have enhanced the application's functionality. Due to its wide acceptance, the initial OBS project has been updated and improved through the efforts of many online collaborators. In its current version, OBS offers users advanced live video production capabilities through a user-friendly interface which can satisfy the needs of most video productions.

OBS allows users the ability to mix media from multiple source types and dynamically manipulate inputs to fit custom projects. The application can handle audio sources, webcams, web browser windows, as well as graphics and much more. In short, it offers a suite of video recording, and live-streaming capabilities that would typically cost thousands of dollars. This is why OBS has become the go-to software for the video recording and live streaming needs of diverse groups of people including small businesses, schools, sports teams, churches, eSports, and more.

What Can You Do with It?

- Record and Stream Live Presentations
  OBS is a perfect tool for recording presentations. These could be recorded PowerPoint presentations, speaker presentations, live meetings, conferences, or church meetings.

- Record and Stream Educational Videos
  These may include tutorials, training videos, and on-boarding sessions for new employees.

- Record and Stream games
  OBS eliminates the need for expensive internal or external capture cards traditionally used for capturing high-definition video. OBS has become the de facto standard for live streaming and recording eSports.

- IP Video Production
  With the newly released plugin support for the NewTek NDI, users can connect to a new world of IP video. NDI allows OBS users the ability to connect multiple OBS systems together. NDI can be used to share video and audio sources over a standard gigabit local area network (LAN).

Who Is It For?

OBS is perfect for use by anyone who wants to make a dynamic video presentation online:

- Schools and Educators: A lot of learning now happens online, and educational organizations are increasingly incorporating blended learning into their offerings to enrich student learning. OBS provides educators and schools with a free and effective means to get started with online teaching.

- Churches: OBS presents an easy way for churches to start streaming services online without having to do expensive system upgrades. With a few tweaks, the software can work with existing equipment to help churches reach new members, shut-ins, and community members on social media networks such as Facebook and YouTube.

- Businesses: Businesses can use OBS to record different kinds of video presentations and stream them to audiences to meet various objectives, including sales, marketing, internal communications, and employee on-boarding.

- Live Events: OBS is suitable for any situation where a person wants to produce a video recording and stream it online, regardless of whether it is a sports event, a conference, or just a reunion.

- Anyone: Even if it is for something as simple as making a video invitation to a weekend hangout, streaming videos from your vacation, or creating videos about your pet. OBS offers the capability to create engaging videos with relative ease.

Where Can I Stream with It?

OBS comes with preset settings for streaming on YouTube, Twitch, and Facebook, but it can be used for pretty much any streaming platform using custom RTMP (Real Time Messaging Protocol). The number of new live streaming networks changes all the time. With an active community of supporters, OBS has remained up-to-date and fully functional as one of the world's most used live streaming software solutions.

2 VIDEO IS THE NEW FRONTIER

Nothing defines today's fast-paced culture more than our impatience, short attention spans, and constant craving for entertainment. In a world where people are looking to get more out of each hour in the day, video has become the preferred medium for communication. Facebook has reported that viewers will spend up to three times as much time watching a live video versus an on-demand video.

People like video for the same reason that children love to have stories read to them—it's passive. Viewers shouldn't have to work to understand your message when it is delivered via video. Just like a great song or story, the author should explain everything to the audience by using his or her actions, tone of voice, and facial expressions. Listening to a story is more engaging than reading by comparison because it is animated for us by the reader ("The Psychology of Storytelling," 2012a).

Before online video came along, there were bedtime stories, songs, and the written word. It's only been less than a decade since the explosion of online video has started to monopolize our culture's attention. The difference today is that we can deliver our stories anywhere, any time, and for as long as we want. And by the way, it is virtually free, assuming you don't mind signing big tech's latest privacy policy.

Video is dominating the internet. Online video fits the modern on-the-go lifestyle. For most people, it is easier to watch and assimilate a ten-minute video presentation than it is to read the same content in text. Cisco notes that by the end of 2019, 80% of all web traffic will be video ("Cisco Visual Networking Index," n.d.-a). Mobile video consumption data from YouTube shows a 100% growth rate in video viewership every year (Ave & Suite 102Clevel, 2018).

About one third of the time people spend online is spent watching videos ("Shifts for 2020," n.d.). Perhaps a few stats will help us better understand the current online landscape:

- On average, people stay on website landing pages for 48 seconds . . . unless it has video—In which case, the average time rises to 5 minutes, 50 seconds (HubSpot, n.d.).

- In 2008, Forrester demonstrated that if a picture was worth a thousand words, then one minute of video was worth as much as 1,800,000 words. Video communicates more powerfully than text and is 5.33 times more effective ("How Video Will Take Over The World," n.d.).

- People are 64% - 85% more likely to buy a video-promoted product, according to Forbes (Bowman, n.d.).

- Recipients are 90% more likely to interact with marketing emails that use video (Hussain, n.d.).

- People's preference for video has made YouTube the second biggest search engine after Google ("Alexa Top 500 Global Sites," n.d.).

What is The Power of Video?

- Video offers an escape from everyday life and lets us momentarily share other people's lives or be part of a fantasy world (Robson, n.d.; "The Psychology of Storytelling," 2012a).

- Video touches more parts of our brain; it combines visual and audio stimuli. Most people identify as visual learners, and since 90% of information reaching the brain is visual, video offers a chance for greater engagement (" | Professional Development for Primary, Secondary & University Educators/Administrators," n.d.).

- Video captures a broader range of emotional experiences when compared to text or other mediums. Humans are emotional beings. People watch to laugh, love, be happy, angry, or shocked, or to relax and learn something novel (Robson, n.d.).

- Video stirs our curiosity. We want to know what the frozen image on a video screen is and to quell that curiosity, we click play and wait for the scene that contains the image ("The Psychology of Storytelling," 2012a).

- Video frees our imagination with content that is idealistic, inspiring, and bizarre—or anything else apart from ordinary ("The Psychology of Storytelling," 2012b).

Ordinary Videos with Extraordinary Impact

One of the best things about online video is that it does not always take big budgets to make a significant impact on a broad audience. As the list below shows, what is required is content that resonates with an audience; something that makes people willing to share the video. People from all walks of life have achieved this and you too can.

- **Video 1: WLR FM Presentation**
  When WLR FM in Waterford City (Ireland) wanted to attract more advertisers to the station, it did not make a long proposal with charts and graphs. Instead, it made a one-minute and forty-five seconds video that explained in clear and compelling fashion why they were the best. It worked! And they became the number one radio station in their area ("(48) WLR FM Is Number One In Waterford City and County! - YouTube," n.d.).

- **Video 2: Little girl Dancing in Church Choir**
  It was little Loren Patterson's first time singing with the children's choir at her church in Dickson County, Tennessee. The

six-year-old was so thrilled; she lost herself in quirky dance moves that someone recorded and put online. It was shared until it went viral. It has now been viewed 23 million times and counting.
- **Video 3: Double Fine Adventure/Kickstarter Pitch**
    A Kickstarter pitch by Tim Schafer for an adventure game his company Double Fine Productions wanted to make. What they needed was $400,000, but what they got was $3,336,371. It was not just the pitch video that made it successful but the idea that the whole game-making process would be on video.

Nicolas Cole a writer who has been featured in TIME, Forbes, Fortune, Business Insider, CNBC and many other major publications offers some timeless advice for content creators. Cole says "So many companies get caught up wanting to see immediate results. I've watched friends go viral (I even had a friend get more views on a piece than I've ever gotten). I've watched clients accumulate hundreds of thousands of views on their work, and get invited to contribute to big publications. It's the people who stick with it the longest, that reap the largest rewards. Don't settle for short-term vanity....I promise you, no good investor thinks that way" (Nicholas, 2018).

Your Turn to Use Video

What can video do for you? If you are not using video to send out your message, you are severely limiting your reach and impact. The move toward video is inevitable and the sooner you learn to employ the medium to its best effect, the better off you will be—regardless of whether you are a small start-up, a not-for-profit, an individual, a church, an educational institution, or an artist.
- Video is perfect for telling stories: A message in text requires work on the part of readers before they can comprehend what you have written. But video incorporates the same nuances of body language that we use in one-on-one interactions.

- Video is adaptable: One video can be tweaked to serve different purposes and speak to diverse audiences. It can act as a salesperson, PR tool, brand spokesperson, or product demo. It is easily adapted to suit style, use, and budget.

- Video is subliminal: It is the perfect way to sell without appearing to. People never like to be sold. But even if a video is selling something, people will watch and share it, as long as it provides them value in some way.

- Video is personal: It gives personality to your business by creating a unique identity. It makes your brand relatable by adding dimension and character to it.

- Video is versatile: It performs well on any device: computers, tablets, or mobile. You can use video on websites, as well as social media. And you can integrate it into an email, pay-per-click, QR codes, and even PowerPoint.

- Video promotes brand recall: 80% of customers remember a video one month after watching it. They also recollect the brand or whatever the video was promoting.

- Video gives you visibility: with video on your website; you can increase the likelihood of ranking on the first page of Google ranking by 53%.

- Video appeals to decision-makers: 65% of business' decision-makers will visit a marketer's website after watching that company's branded video. 59% of company decision-makers prefer to watch a video, in place of reading an article, proposal, or blog. What this means is that online video is now the fastest way to share information around the world. This trend is not likely to change soon. Videos are shared around 1,200% more than both links and text combined. (Clarine, 2016; Guerin, n.d.)

Getting It Right with Video

Since video offers such potential for helping to get your message out to the most people, doing video right is crucial. To succeed, you need to find a combination of the right message, colors, music, tone, style, and marketing strategy. Making your video is just one part of the online video equation; getting it out is the other part. This book addresses vital aspects of creating live video and streaming your content to the world. Reading this book will help you understand how to use one of the most cost-effective software solutions for recording and live-streaming video.

I hope that this book can help put the skills and technologies that organizations used to pay thousands of dollars for at the fingertips of just about anybody. If you follow the topics discussed in the next pages attentively and implement them, you will overcome many of the obstacles people encounter trying to setup live streaming and video production systems with OBS.

And best of all, as already mentioned, the software is free to download and use. So, you have no more excuses for getting started even with the most basic of projects.

So, let's get going!

3 THE LIVE STREAMING LANDSCAPE

Steaming is how much of the information you access on your internet-enabled devices reaches you. "Streaming Media" is a broad term used to describe technology that transmits data to your computer and mobile devices on the internet, whether it be, music, video, or live communications. When content is streamed, it is possible for a user to access it immediately from an online server without having access to that file locally on the device.

Not everything we get off the internet is streamed. Downloading content from the internet is not considered streaming. That is because when you download software, for instance, from the internet, you cannot start to use it until the download is complete. But when you click play on a music app, the song begins to play immediately. You do not need to wait for all of it to download before you listen because the content is streamed as needed.

In the past, most digital files you had access to on the internet were only available via a progressive download. For a long time, you had to download music and video content directly to your computer if you wanted to use them. In the last decade, however, advances in encoding, compression, and bandwidth access have made streaming content commonplace. Most people do not feel the need to store the music they love on their phones anymore, because they can easily access these files from the cloud. The same is true for videos on YouTube. Viewers do not need to store data locally on their devices. They are always available for streaming whenever users log into the platform.

So, what if you wanted an audience to access an event in real-time, like watching a live football match, what would that be called? That is "Live Streaming." Live streaming is content that is delivered as it happens, to an audience that connects through the internet. It has been popular with talk shows and news broadcast for years on television. The difference today is that anyone can now live stream. Today live streaming is used for webinars, Q&As, sports events, church services, business presentations, and everyday video gaming.

Live Streaming

When video content is recorded and broadcast simultaneously over the internet so that viewers can see the action as it unfolds, you are talking about live streaming. Live streaming video is highly diverse and applicable across a variety of scenarios. Users can adapt how they use it and make it do just what they need. This flexibility is a significant reason for its quick rate of adoption ("Live Video Streaming," n.d.).

The Attraction of Live Streaming

Live streaming is exploding on social media as a new form of authentic communication. Neil Patel, a famous online marketer says, "It's one of the most genuine ways to connect with an audience and allows for levels of personalization that the marketing industry has never seen" ("Why You Should Care About Live Streaming in 2018," 2017). The same things that make Reality TV popular are also at play in live streaming. But, unlike Reality TV, live streaming is not rehearsed, censored, or edited before delivery. While many "live streamers" focus on highly professionally produced content, various new, unedited forms of "in real-life" live streaming are becoming more popular. The immediacy and excitement of live streaming video make the medium a big draw for audiences worldwide ("Yahoo_The Live Video Opportunity_2016.pdf," n.d.).
Here are just a few of the factors that make live streaming so powerful:

- Exclusivity: Live streaming offers viewers a feeling of being part of a select group. New forms of premium access to behind-the-scenes action is creating many profitable opportunities on the internet.

- Capture Attention: A live stream audience is less likely to get bored and leave. Live streams can hold audience attention for hours versus seconds and minutes, like other kinds of video.

- <u>Time-Bound</u>: There is a sense of urgency associated with a live stream. Viewers know if they do not watch, they will be missing something that an active audience chat room is enjoying.

- <u>Community Building</u>: Broadcasters can see and reference comments in real time, allowing them to identify individuals by name.

- <u>Authenticity</u>: Because it is not edited and cannot be rehearsed, live streams have become a way to genuinely connect with an audience and create a more authentic experience.

- <u>Control</u>: It offers the audience a great deal of power because they can determine the flow of the interaction; they also have the chance to dictate the direction of future content.

- <u>Affordable and Easy</u>: Creating a live broadcast no longer requires expensive equipment and software expertise. With a webcam, an internet connection and OBS, anyone can use their Mac, PC, or Linux computer for live streaming.

Live Streaming Platforms or CDN

A live streaming platform or "CDN" (Content Delivery Network) is where you send the live feed of your video and generally where the audience goes to view it. The biggest CDNs include Facebook, Instagram, Twitter, Twitch, and YouTube. For good reason, most people stream to social media networks where they already have a substantial presence. In this way, content creators can grow their subscriber networks and leverage social media notifications on each platform as a way of organic communication with their followers.

There are a multitude of other smaller—and sometimes niche—platforms including Vimeo, IBM Cloud Video, DaCast, and StreamShark, to name a few. In this book, our focus will be primarily on the top free live streaming platforms: YouTube Live, Facebook Live, Periscope, Twitch, and Instagram Live.

Let's review some of the basic details you should know about the top CDNs available today.

Facebook Live

Facebook Live initially launched in August of 2015 to a privileged group of users before it was rolled out to the broader Facebook community in April of 2016. Below are the most essential features of the service:

- Everyone on Facebook can go live.

- It has real-time interaction and streaming notifications.

- It's easily accessible on mobile devices and desktops alike.

- Streaming content is available on-demand with the live chat room comments displayed.

- Detailed analytics are provided about viewers and broadcasts.

- Exclusive live streams are available inside Facebook groups.

Periscope

This streaming app was developed for Android and iOS and acquired by Twitter in 2015 before its launch. It gained 10 million users in the first four months and hosted 200 million broadcasts the same year. Periscope Producer now allows Twitter users the ability to live stream directly to Twitter. Its important features are:

- Live streaming direct to Twitter via Periscope Producer.

- Incorporates notifications, chat function, and social sharing.

- Replays are available for 24 hours after event.

- No restrictions on length of stream.

- Popular with the age 16 to 34 demographic.

YouTube Live

YouTube launched the live streaming revolution with a live feed of a conversation with President Barack Obama back in 2010. The platform is known as one of the most advanced and user-friendly streaming services available today. More recently, YouTube started to allow channels to live stream from the mobile app. Below are features of YouTube Live:

- Streaming made available via custom RTMP and a mobile app.

- Videos are archived to be used later.

- Streamers share ad revenues with the platform.

- Highest quality 4K video streaming.

- Live streaming channel page with static streaming key.

- No stream length restrictions.

- Low-latency live streaming delivery options.

Instagram Live

Instagram Live allows users to stream live content to their followers on the platform. It combines some of the features that users are familiar with from Facebook Live:

- Instagram has been designed to be a mobile-first platform.

- Seamless integration with Instagram stories.

- Replays are available for 24 hours after event.

- Attracts a younger demographic between ages 18 to 34.

- Great for Q&As and more flexible content.

- Hair, makeup, travel and entertainment content thrive on this channel.

Twitch

Twitch is an online live streaming website owned by Amazon. Amazon purchased the popular Justin.TV live streaming platform and combine it with Twitch.TV. Twitch is now one of the most popular and interactive live streaming destinations especially for eSports.

- Twitch includes low latency live streaming features for eSports and higher engagement.

- Twitch includes new extensions developers can create and streamers can use. The extensions provide new interactive on-screen overlay features that extend the broadcasters interactivity capabilities.

- Twitch includes a monetary system called bits used for audience members to tip broadcasters.

- Twitch is a live streaming first platform and has seen record numbers of watch time rivaled only by YouTube.

The Future of Live Streaming

As big as online video is currently, it is only expected to get bigger. And live streaming is going to be a considerable part of that growth. Facebook's CEO said in 2016 that he believes the platform will be almost all-video in five years ("Mark Zuckerberg: Within Five Years, Facebook Will Be Mostly Video | Popular Science," n.d.). Video is going to become even more ubiquitous over time. Brands, as well as individuals, will get more creative and adventurous in their use of the medium.

Here is a summary of where live streaming is today and the direction it is expected to grow.

Current Trends

These are the uses of live streaming that have become dominant since the technology became accessible to everyone:

- Soaring popularity of live video clips as a means of communication among young people.

- Social media platforms accelerate their movement into the live streaming space.

- Live sports events broadcast to record-breaking audience sizes.

- Live streaming of eSports and video game play become immensely popular.

- Journalists use live video to cover current events from remote locations around the world.

- Houses of worship adopt live streaming to connect with distributed worshippers.

- Live video gains ground as an essential marketing tool for brands.

- Influencers use live video to connect and stay in touch with followers.

- Live video becomes an accepted medium for companies' internal communications.

- Educational institutions start to merge live video with more traditional forms of teaching.

Future Trends

No one can predict the future, but based on where things have been, and where they are now, we expect the future of live streaming to be something like what a Cisco White Paper on the subject predicts ("Cisco Visual Networking Index," n.d.-b; "The future of live streaming video," 2018):

- Live video is going to become as common as a facetime or video call.

- Just as print and video, in their own times, became the lynchpin of corporate marketing, live video is expected to become the central tool brands use to educate, inform, and connect with their customers.

- Independent news reporting accompanied by live videos will be made available from various locations around the world. Many viewers will turn to these sources for news because they will offer more authenticity and local authority than traditional news channels.

What Does It Mean for You?

It means you should act now. There are boundless opportunities available to organizations, small and large, with low barriers to entry. Regardless of whether you want to:

- Increase your personal or business brands followers on social media.

- Help a social cause that is close to your heart.

- Find new clients to hear your business proposal.

- Become an indie news reporter.

- Promote a local sports event or concert.

All of these can be achieved via live video. And to do that, we are going to prepare you with a course on Open Broadcast Software—**The Unofficial OBS Live Streaming Course**. In the busy world of the internet, video has become an essential tool for accomplishing many of these goals.

This is why you should invest the time it will take to learn the best techniques from our team here at StreamGeeks. When the time comes, you will be well prepared to profit from the opportunities of working with video.

The rest of this book will discuss how to use OBS to become a live video producer. This book includes an accompanying online course: (**The Unofficial OBS Live Streaming Course**), with lecture sessions designed to equip you with the little tricks and skills that will separate an average video from a phenomenal one.

Do not forget to reference the included glossary at the back of this book as needed. This book should help you pick up the vital vocabulary often used in live video production. Understanding the language of streaming media will be the foundation you can use to help achieve your next live streaming projects goals.

Now, on to the next chapter!

4 OBS SOFTWARE COURSE OVERVIEW

The StreamGeeks OBS live streaming course is one of the few courses of its kind on the internet. Most OBS courses have been designed for video gaming, which is the main reason Hugh "Jim" Bailey originally created OBS. This book and online tutorial course caters toward the needs of regular people, businesses, and volunteers.

You can find the course here - https://www.udemy.com/obs-live-streaming-course

What you will learn in this course

At the end of this four and a half hour, self-paced video course, you will have a better understanding of:

- The skills needed to create professional live streams and video recordings using Open Broadcaster Software.

- How to stream live to Facebook or YouTube

- How to create an information ticker. The horizontal text that usually crawls along the bottom of TV screens, especially during the news.

- Adding annotations. How to use texts, comments, and emojis that often appear over a video to help explain a point.

- How to work with all kinds of media inputs. Learn how to apply filters and creatively mix together media.

- How to set up your audio to sync perfectly with the video using a tool that gives accuracy up to a hundredth of a second.

- Tools used to enhance your voice and make it sound crystal clear in your recordings.

- How to use a virtual set—a background you can manipulate and change to suit whatever you need it to be.

- How to use Adobe Photoshop and After Effects with the included course files to create a fully branded live video production setup.

The course is designed to take you from OBS "newbie" to OBS rock-star in record time. With this book you can go "*zero to 60*" with your OBS knowledge as fast as possible. This is the ideal place to start your learning journey if you are serious about using OBS for video production and live streaming.

Who Is the Course For?

This course has been created with the needs of "non-techie" people in mind. It is ideal if you are:
- A business
- A house of worship
- An educational institution
- A Not-For-Profit
- A film or video production studio
- Or a hobbyist

Course Prerequisites

- A Mac, PC, or Linux-based computer.
- A YouTube or Facebook account (For live stream testing).
- Hard drive space for the necessary practice materials that are included with the course and the OBS software.
- **Optional**: To create custom graphics and animations, you will need Adobe Photoshop and After Effects (or similar editing and animation software).

Section ONE

Section 1, Lesson 1: Introduction to the course

Guess what? This course has been updated since its original publication in 2017. The entire course has been re-recorded and updated to provide answers to the most critical questions students have been asking about OBS. Please feel free to use the Udemy online communication tools to post questions and get answers.

Section 1, Lecture 2: Downloading OBS

To get started with OBS, you have to download the software. That is easy; enter "Open Broadcaster Software" in Google, and it should be the very first link on the resulting page. Clicking on the link will take you to the download page for OBS.

Also, you may follow this link https://obsproject.com/. Copy and paste it into your browser to be taken to the main OBS project website.

There are three download options for the different operating systems: Windows, Mac, and Linux. If you use Windows, be sure to select the 32-bit or 64-bit version, depending on your system. You can check to determine whether you have a 32 or 64 bit system by searching in Windows for "About My PC." There is also an option to download the classic version of OBS, but you should know that this version is no longer supported. Furthermore, you may also choose to download an older of OBS if you do not want to use the latest version, which is what you will download by default. Click the download button in the top menu and, in the page that opens, follow the link that says "Previous Releases" to choose your preferred version.

Finally, the social media links to OBS social media pages are prominently displayed on the website, and it is a good idea to like and follow them on Facebook, Twitter, and other platforms to stay up to date with new releases. It's worth checking out the OBS community on the forum page as it offers tons of opportunities to benefit from the knowledge and experiences of other users. Note that the OBS forum is where you will find all the plugins for OBS.

Section 1, Lesson 3: OBS Interface Overview

In this lecture, you will walk through the OBS interface and learn some of the settings you will need to apply to get the best performance out of OBS.

The OBS interface has been designed to give users a straightforward step by step process for building dynamic live video productions. First, don't be surprised if the OBS version that you have doesn't look exactly like the one pictured shown in this book. It's worth checking if there is an updated version of OBS by clicking the "**Help**" drop-down menu and selecting "**Check for Updates**."

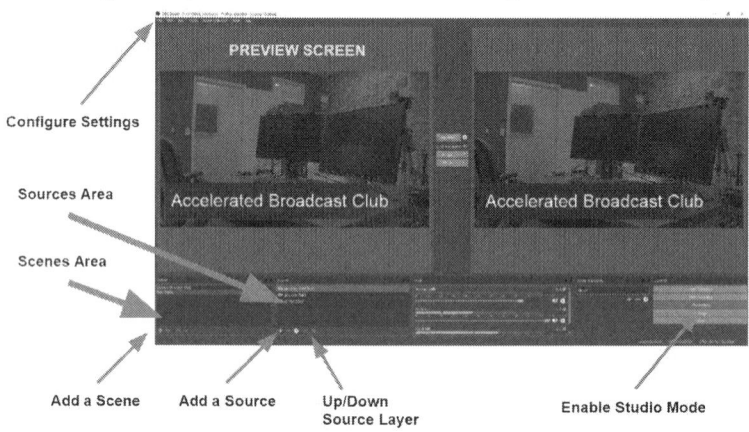

The latest version of OBS now includes a couple of nifty new features you will want to start using. Start by clicking **"Studio Mode"** in the bottom right hand corner of OBS, to bring up a preview screen. You can use studio mode to transition between multiple scenes and configurations in your setup. If you like the darker gray colored theme used here, you can enable that in the settings menu under **"Themes."**

The main inputs inside OBS are called **"Sources."** Each source is organized inside a **"Scene."** You can start your OBS project by naming a scene and adding sources into that scene. In this way, you can build unique media layouts and transition between them throughout your broadcast. Each scene and source can have a unique name for organizational purposes. You can name your first scene by right clicking it and selecting the **"Rename"** option.

Next to the scenes area is your sources area. You can add an input using a plus button (+) and select the type of media you would like to add. Each source can be arranged in a system of layers. You will find up and down arrows in the sources area which are used to organize your sources position in the layering system. Each source is arranged top to bottom so that you can display your media in a stack. You may want to show animated graphics, on top of a live camera, with another piece of media on top of that for your logo. Using the up and down arrows, you can determine the order in which your media will be stacked together inside your scene.

OBS in Studio Mode

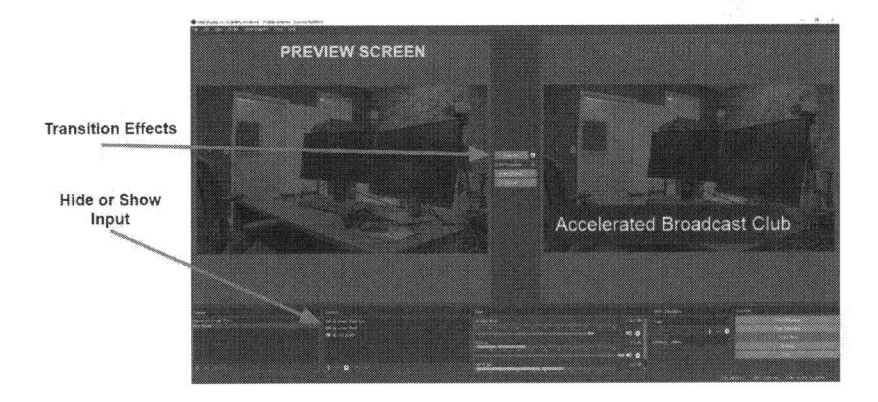

Using **"Studio Mode"** you can build a completely redesigned layout of your current scene by hiding or revealing new elements in the sources area and transitioning to the new layout. You can adjust a scene in the preview area before it is sent to the live output screen with a transition button. Once you have changed the scene by adding or removing a lower third or another type of media, you can select a transition type to switch your preview screen content to the output of your recording or live stream. In studio mode, the left-hand screen is always a preview area, and the right-hand screen is your output.

Optionally, you can build different scenes before your production starts and switch to each throughout your broadcast. A popular way to use OBS is to turn on an option in the settings area to make your scenes **"go live with a single click."** In this setup, you don't need to enable studio mode at all. You can click (or use a hotkey) to switch between your scenes on the fly. The drawback to this setup is that you cannot select custom transitions in-between your scenes. There is a new feature called **"Transition Override"** that can be used to override the default transition for each scene.

Putting it all together in OBS

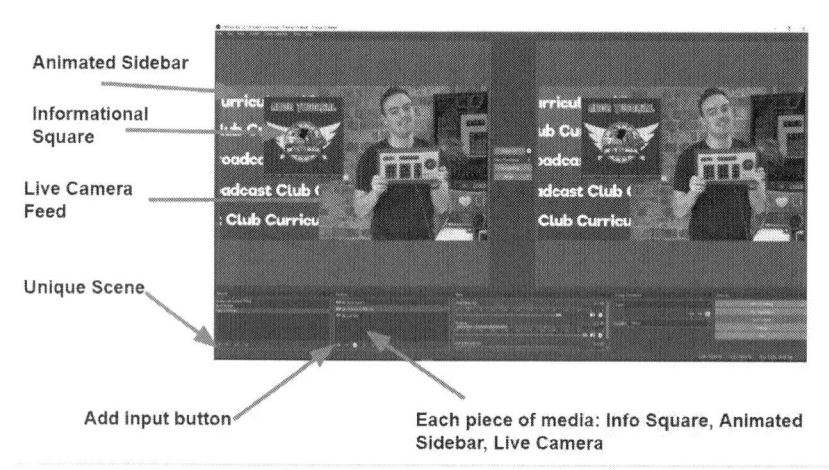

In the above picture, you can see a sample media stack we use at StreamGeeks for our live productions. We have a live video camera feed for our base layer with an animated sidebar used with an informational square placed above it. The animated sidebar is a video that has been set to play on a loop.

The StreamGeeks producer will sometimes have multiple informational squares which can be revealed to show new content. In this way, you can present supporting information for our presentation that is made in the way you might see on live television. In this layout, you can use a single scene with multiple layers of media that are stacked on top of each other. This way, your producer can hide a square of information which will reveal a new square that is positioned directly under it.

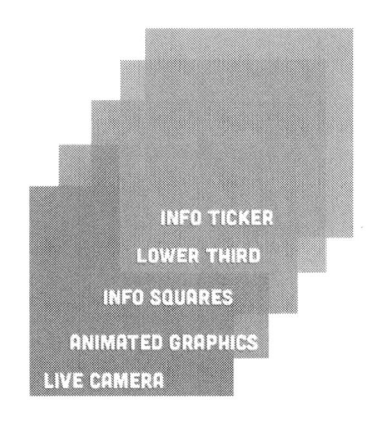

The sources menus has a cog that can be clicked to take you into the specific settings area for the source you have selected. You can create as many scenes as you need and add as many sources as you need inside each scene. In the example, we have four scenes: the base scene, a scene with ticker, a chat room, and a split screen. On the "Scene with the Ticker," for example, there are four camera sources. But "The Base scene" shows only one NDI camera as its source. The scenes and sources menus give you the flexibility to build scenes and add sources as you want.

In the settings area, click the **"Stream"** tab on the left. You can use this to configure your streaming settings, which we will cover in more detail later. For now, familiarize yourself with the options and go to the "Output" tab. You will see that your "**Output Mode**" is set to simple. Simple mode is not very useful for this course, because it offers a limited set of options for controlling the output. For this course, you should use the advanced output option, which gives you the full set of controls. You will gain a basic understanding of how these settings work throughout this book.

On the left pane of the settings window, move to the "**Audio**" tab and click on it to access the default sample rates for audio. Start to familiarize yourself with these settings as they may become necessary for troubleshooting in the future. The "**Video**" tab can be used to see your video resolution options. We will go over the optimal settings for both audio and video later on, including the "**Hotkeys**" area which includes advanced settings.

Click "**Okay**" to close the settings window and click the "**Edit**" dropdown menu. Look for "**Advanced Audio Properties**" in the dropdown menu and click on it. You can also access this area from the audio mixer on the main OBS dashboard using the cog icon. Advanced Audio Properties allows you to decide which audio sources you want to send to which track and you can use the "Sync Offset" feature to sync your audio and video. This will be covered in more detail in the following chapters.

You may also want to customize the appearance of the OBS display by dragging and dropping the different controls on the main screen to wherever you want them. Using the "**Scenes Collections**" tab on the top menu, you can also create different collection of scenes for various purposes. This will come in handy if you have different versions of one presentation for different types of shows or live streaming scenarios.

Finally, the "**Tools**" tab on the top menu gives you access to additional tools which will sometimes include features from installed plugins. That is all for the interface; most of these settings will be explained in detail in subsequent chapters.

Section 1, Lecture 4: Zero to Sixty

Let's open up OBS with a completely blank slate with no saved scenes or sources. The goal will be to build out some commonly used scenes for video production. You can start with a basic scene that includes a single camera, another scene with an animated ticker, a scene with an interactive chat room from YouTube, a split-screen presentation, and a full-screen presentation layout. To begin, create a new scene collection by clicking "**New Scene Collection**" on the top menu and selecting "**New**." In the example, you can rename it "Zero to Sixty" but you can call yours whatever you want.

Start by renaming the default scene as the "Base Scene" and bring your camera source into the scene by adding a webcam source. The example used in our video is a NewTek NDI source which you will learn more about in an upcoming chapter. After you have done this, right-click on the base scene to duplicate it and rename the new scene to whatever you want it to be. The one in the example is renamed 'Scene with Ticker.' This will completely duplicate the scene and its sources. You can remove the cameras that you don't need and start customizing the new scene.

Adobe Photoshop is a great tool for creating media that can be used to brand your production. In this tutorial, Photoshop is used to modify the files used for this setup. Remember that you can download these files from the Dropbox link provided in the resources section of the online course. Just click on the hamburger icon on the top left corner of any of the videos—the table of content and resources section will open. Under the first lecture, you will find the link that says "OBS Course Files;" follow that to download the media files for the tutorial.

The Photoshop tutorials are included to give you an idea of how to build these media files yourself. Once inside Photoshop, choose a 1920X1080 space and widescreen 16:9 setup. To create a custom lower third, we will draw a rectangular strip at the bottom of the canvas. You can paint it black, add an accent color, and save it as a PNG file to import into OBS. Next make you can optionally make a mask file, into which you can add your logo. Now we will add an image into our scene: click sources, select image, and rename the image to "Ticker Background." From the file browser window that opens, select the ticker background image you downloaded and apply it.

Next, you can add another image, which will be your 'Ticker Top' and for that we select the image called "Ticker Mask Layer." (Note: if this is confusing to read, consider reading along with the video tutorial). This will put a layer on top of the first image, creating a mask effect. After this, you can add some text and call that "The Ticker Text." You do not have to use the same text as the example. Copy and paste the text several times, so that it repeats or write out custom text. Select color, font, and font size for the text, and that is it. After you click "Okay," you may have to go back and readjust the size and font, if they do not fit inside your ticker.

To make the words scroll across the bottom of your screen, which is what a Ticker is supposed to do, right-click on the ticker text in the sources pane and select "**Filters**." Inside filters, there will be a little plus button at the bottom of the window. A list of effects will appear, and you should select "**Scroll**." You may now choose how fast you want the words to scroll by dragging on the bar. If the scroll speed is too fast or slow, go back and adjust it until you get the right speed.

Next, you can set up a scene with an integrated chat room. Start by right clicking on the base scene to duplicate it. Rename it to "Video with Chat" for this example. Remove the camera sources that you do not need and then head over to Photoshop. Start by creating a file size that matches your settings in OBS. For this example, use a 1920 by 1080 pixel space. Inside Photoshop, you can start to plan out your graphical layout and create a large box on the left where the video will be overlaid. On the right side, leave space for the chat room, which will be displayed using a web-browser source. Once you have created this layout save the file as a .png with the name "Chat Plus Video." Now back inside OBS, you can add the image called "Chat Plus Video."

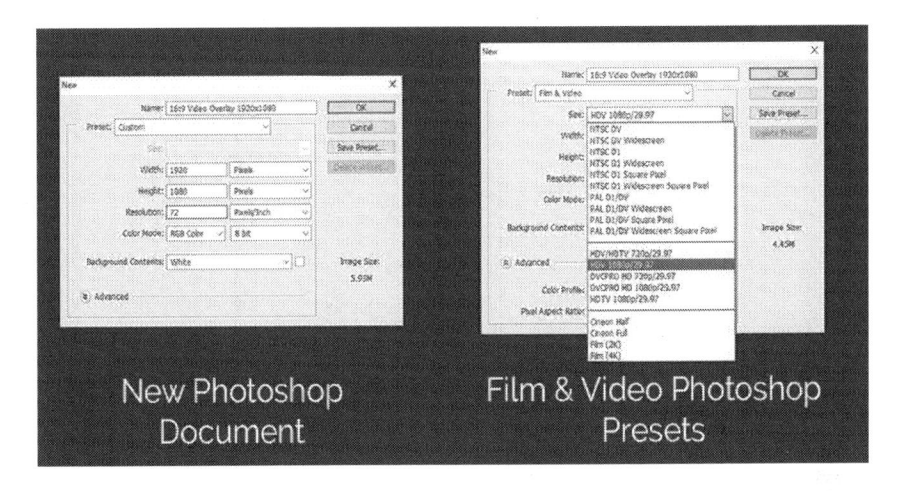

New Photoshop Document

Film & Video Photoshop Presets

Once your image has been inserted in your scene, it's time to set up your chat-room. Go over to YouTube and, inside your account, go to the live streaming area. On the right side of your channels live streaming dashboard, look for the chatroom and click on the menu on the right to pop it out. Copy the link inside the popup window. You can now go back into OBS, and choose your source as "Browser input." Go ahead and paste the link you just copied into the box and then select the size you want your chat room display to be. That will put the live chat room on the screen right beside the live video. This is an excellent example of using the web-browser input feature inside OBS. You can bring any website into your video production space in almost any size.

Next, let's duplicate the base scene again, and rename it as a presentation scene. You can click the "**Sources**" plus button to choose the image option, and then select the layout image that you want. You will find some example images in the course downloads on Udemy. After that, you can choose a camera input or screen capture area to be your presentation. To make this scene look professional, make sure it fits into the presentation area by dragging the pin at the edge and resizing the entire area. On the left-hand side of our presentation area, where there are two bubbles, you can insert live cameras into which will show yourself during the presentation. That way the large square shows the material you are discussing and the smaller bubbles show the presenters.

> Pro Video Tip
>
> Don't forget to have B-Roll video ready to play during your broadcast. It's interesting for your viewers, but it also gives your talent a moment to prepare off camera.

Finally, you can set up a split screen scene for use in many scenarios. Again, duplicate the base scene and rename it as "Split Screen." Select sources, choose image, name the image, and then go to the course files and select the image "Split Screen" from the split screen folder. Once you insert that, and after you adjust your video a bit, you should have a split screen. Having options during your broadcast is a great thing. If you are learning how to use OBS, these simple tutorials are a great way to exercise your capabilities.

And that is it with the "Zero to Sixty" presentation . . . In sixteen minutes, you have seen that OBS is not that difficult to use.

Section 1, Lecture 5: How to Optimize OBS for Streaming & Recording

When it comes to understanding the recording and streaming options in OBS, it's best to get out of "**Simple**" mode and into the "**Advanced**" menu area. To get started, navigate to the "**Settings**" area by clicking on the "**File**" tab in the menu bar or in the lower right-hand side button area. This will open a window with all the main settings for the software. By default, the settings bar opens in the general settings; other settings can be accessed on the left side of the window.

Video Recordings in OBS

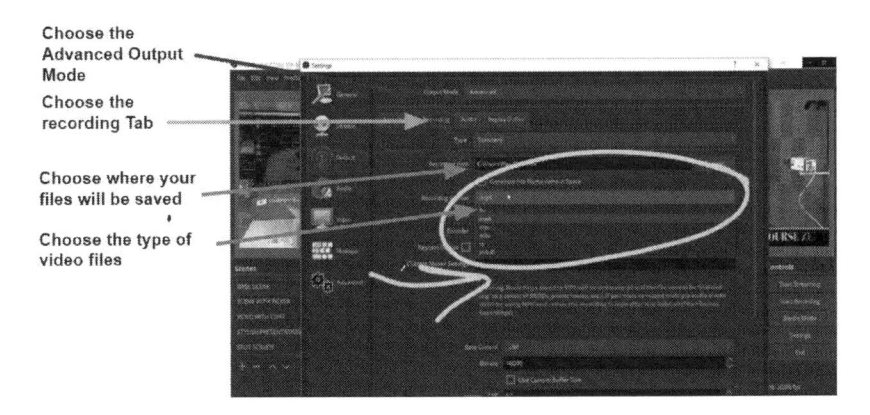

Choose the Advanced Output Mode

Choose the recording Tab

Choose where your files will be saved

Choose the type of video files

Under the "**General**" tab, you can have OBS automatically check for updates. You may also choose to show a confirmation notice when you start streaming. Here you will also find settings for how you want your projector to behave if you are using one. You can set up your projector output to always be on top of other windows, and you can hide the mouse cursor. You can set up how you want your "**Studio Mode**" to switch between scenes. There is also a new section with options for the Multiview, which will be discussed in a later chapter.

When it comes to live streaming and recording one essential tool, you should keep an eye on is your Windows Task Manager (Activity Monitor for Mac). When you are recording and streaming video, this process is called "encoding." Your computer is encoding your OBS production into the video format of your choice and then streaming or recording this

Pro Video Tip

If you are using a Windows laptop for live streaming with OBS. Check your power settings and make sure you are using the "Performance Mode."

information to the destination of your choice. This process can take up a lot of processing power. A tool like Task Manager will help you determine how much processing power OBS is using and how much bandwidth you are using. You should periodically check the bottom right hand side of OBS to view the CPU usage being reported.

When you are choosing settings for video streaming and recording the first choice you need to make is resolution and frame rate. Once you have determined your project resolution and frame rate, you can choose a bit rate. The bit rate is essentially the quality of your recording inside the resolution canvas. A 6Mbps bit rate live streamed in 1080p is considered excellent quality by today's standards. When you are recording to a local hard drive, you can generally use a higher bit rate because your upload speeds do not restrict you.

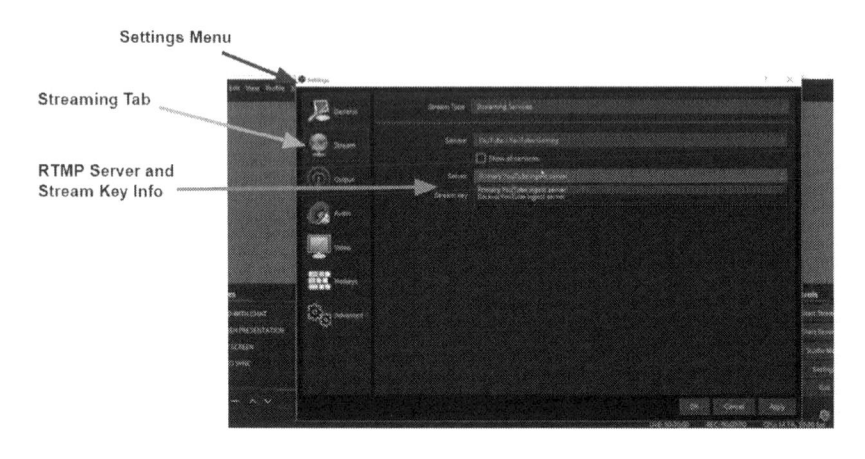

When you are live streaming, a good rule of thumb to follow is to only use half of your available upload speeds for live streaming. When you are recording locally to your hard drive, you can choose a higher bit rate from 1-100Mbps. 100Mbps is considered high quality when recording in 4K with a high-end camcorder. For 1080p recordings meant for YouTube or Facebook, I generally use 8-12Mbps. The higher the bit rate, the higher the video quality. Most video makers find a happy middle ground between the hard drive space required for higher bit rates and the quality.

> **Pro Video Tip**
>
> Always try to schedule your live broadcasts. On YouTube specifically it helps generate more watch time for your on-demand videos and increases organic notifications.

The first thing about live streaming that you will need to know is where you can retrieve your CDN's RTMP information. A CDN is a "Content Delivery Network." Facebook and YouTube are both CDNs which provide RTMP information in the form of a server name and a secret key. The **"Stream"** tab is where you can enter the RTMP information for OBS to connect to your preferred content delivery network. You can select from a list of known live streaming server addresses using the dropdown menu or use a completely custom RTMP URL and key provided by a CDN. When given the option, always choose the server that is closest to you to avoid delays.

If you are starting to learn about bandwidth and video storage, it is important to remember megabytes are used for files on a hard drive and megabits are used for streaming data on the internet. In my opinion, streaming in SD is no longer acceptable, and we must understand the bandwidth needed to stream in HD. Generally, the minimum resolution you want to live stream an event in would be 1280x720p with a 1.5 Mbps bit rate. 720p resolutions are technically considered "High Definition" but remember that the bit rate is the real measure of quality when we are talking about streaming video.

Resolution	Pixel Count	Frame Rate	Quality	Bandwidth
4K 30fps	3840x2160	30fps	High	30Mbps
4K 30fps	3840x2160	30fps	Medium	20Mbps
4K 30fps	3840x2160	30fps	Low	10Mbps
1080p60fps	1920x1080	60fps	High	12Mbps
1080p60fps	1920x1080	60fps	Medium	9Mbps
1080p60fps	1920x1080	60fps	Low	6Mbps
1080p30fps	1920x1080	30fps	High	6Mbps
1080p30fps	1920x1080	30fps	Medium	4.5Mbps
1080p30fps	1920x1080	30fps	Low	3Mbps
720p30fps	1280x720	30fps	High	3.5Mbps
720p30fps	1280x720	30fps	Medium	2.5Mbps
720p30fps	1280x720	30fps	Low	1.5Mbps

The chart above displays various bandwidth choices you will have for your live streams. Using this chart and your available upload speeds, you should be able to map out the quality of live stream that your internet connection can support. Now might be a good time to perform a free bandwidth speed test. Simply Google "Bandwidth Speed Test" and Google can perform the test right there!

On the "**Output**" tab, you have an option between simple and advanced settings. With the simple output mode, you can choose your bit rate (to determine the quality of your live stream), your encode option, and the audio bit rate. The output section is also used to manage your recordings. You can choose where you want the recording to be saved, the quality you want it to be, and the file format you prefer.

The advanced output mode under the "**Output**" tab allows you to dig into the full functionality of OBS. Choosing the correct encoder setting can significantly increase the performance of OBS, especially if you have a graphics card available on your computer. You will find x264 is the standard encoding option on most computers. It's worth testing out the other encoding settings that OBS automatically finds because one of them will likely leverage an on-board graphics card which will free up CPU processing power for OBS. You will also get the option to rescale video outputs to match the platform you are streaming on. Keep in mind that rescaling your encoded video stream or recording will take up additional processing power on your computer, and it is generally not recommended. In advanced mode, you gain access to an audio matrix which allows you to select one audio track for your encoded video stream and multiple audio tracks for your recordings. Audio tracks are managed in the "**Advanced Audio Properties**" area and reviewed in more detail in a later chapter.

In the "**Audio**" tab, you have settings for choosing a sample rate for each audio track. The best settings for your sample rate will depend on the input from your audio devices. Most devices use a sample rate of 44.1khz. In the audio tab, you can set up all of your default audio devices. These are the devices that will show up in your audio mixer, which is shown in the bottom center area of OBS by default.

In the "**Video**" tab, you can set up your project's resolution size and frame rate. You will notice that the two default dropdown menu options are 1920x1080 and 1280x720. These are the most commonly used formats, but you can enter any pixel resolution size that you want to. This is a great way to create square or mobile video content. For example, you may want to create a 1080x1080 square canvas size. Or you may want to create a 1080x1920 portrait video stream to Facebook. It's essential to think about the entire workflow of your video equipment. If your video equipment is set to 1080p at 30 frames per second (or 25 fps), you should set up your OBS canvas to match your equipment. Most users will set the frames per second at 30fps, but for high motion applications such as sports, 60fps (or 50 fps) is ideal. Keep in mind that if you double your frame rate, you are also doubling the bandwidth required to stream your content.

Finally, you will notice "**Hotkeys**" and "**Advanced**" sections in the settings menu that will be covered in a later chapter. At this point, you have just opened the hood of OBS and reviewed how to configure the system for your project. It's now time to get started with some real-world applications.

Section TWO

Section 2, Lecture 7: OBS Sources & Filters Overview

OBS offers a long list of potential input sources that you can use to create dynamic live video productions. By default, OBS will open with a scene creatively named "scene" which you can rename by right clicking on it. By this point, you know that you can click the plus button at the bottom of the "**Scenes**" box to add additional scenes. OBS has a handy feature called "**Scene Collections**" which allows users to create collections of scenes and load them into an OBS project. For example, you can have one production with scene A, B, and C, and then you can have an entirely different collection you can load into OBS with scenes E, F, and G.

Tip: You can duplicate your favorite OBS scene collection and use it as a starting point for creative projects. This way, you can always go back to the base scene collection if need be.

As of OBS 23, the following input types are available to users: Audio Input Capture, Audio Output Capture, Browser, Color Source, Display Capture, Game Capture, Image, Image Slideshow, Media Source, Scene, Text, VLC Video Source, Video Capture Device, and Window Capture. The following list explains each source in detail:

- Audio Input Capture—This is used to add audio sources available to your operating system. Audio setup is currently best handled in the "**Audio**" section of the "**Settings**" menu.

- Audio Output Capture—Your computer likely has multiple audio outputs. You can bring these audio streams into your OBS production with this source.
- Browser—This source allows users to bring any webpage into their video production. Simply type the URL of the webpage you would like to add into your production and press OK. This input also allows users to apply custom width, height, frames per second, and CSS.
- Color Source—The color source input allows users to bring a custom color layer into their production with any width and height. This is useful as a base layer in many productions.
- Display Capture—Display capture is used for capturing the content from a display connected to your computer and bringing that video into your production. This is ideal for PowerPoint slides and other non-gaming uses.
- Game Capture—Game capture is used for high-frame-rate gameplay capture. This source will reduce latency in comparison to display capture.
- Image—This input allows users to bring in media that is in the formats of bmp, tga, png, jpg, or gif.
- Image Slideshow—The image slideshow source allows users to bring in a selection of images into their production. Image slideshows can be setup to loop automatically or transitioned with hotkeys.
- Media Source—Media files can be brought into your OBS production with this source. Media source supports the following formats: mp4, ts, mov, flv, mkv, avi, mp3, ogg, acc, wav, gif, and webm.
- Scene—The OBS scene source allows you to bring an entire scene within OBS into another scene.

- Text—This source allows users to bring text into their video production space. This input includes options for font, color, size, and more. Filters can be applied to this source to make the text scroll.
- VLC Video Source—This source allows users to bring video and media directly into OBS using the VLC media player. This is ideal for RTSP video streams and other types of advanced media.
- Video Capture Device—The video capture device source is used webcams or other "video capture" devices like an HDMI to USB frame grabber. Other video capture devices used with this source include PCIe capture cards.
- Window Capture—The window capture source allows users to capture content from a specific window open in their operating system.

Note that you have to add each source that you want to include in any scene manually, unless you have duplicated an existing scene. Just because you have, for example, five cameras and three microphones connected to OBS, it does not mean that they will be automatically inserted to every scene. When adding a source, you will see a list of other sources that are already existing inside OBS. You can add sources that are already existing in different scenes to as many scenes as you would like within reason. Staying organized is essential, so choose a name that will clearly distinguish the source from all other sources you plan to use.

As you bring in sources that include audio, you will notice that they are added to the audio mixer automatically. You can mix these sources by clicking on the cog in the mixer box. You can also mix each input separately by clicking the cog on that input. And that is it for how to bring sources into OBS.

Any source can have a selection of two types filters applied to the source. These filters can be used to produce visual effects or enhance the sources presence in the overall production. You can choose from "**Audio/Video**" filters and "**Effect Filters**." Here is a list of filters available by default in OBS:

- Audio/Video Filters

    - **Compressor** – A compressor can be used to reduce the level of an audio signal when it hits a certain ratio and threshold. Compressors can make your audio sound better and more uniform especially if speakers tend to get excited and yell.

    - **Expander** – An expander can be used to increase the dynamic range of your audio sources. Expanders are like a noise gate and can be used to a subtler effect.

    - **Gain** – Gain can be used to increase your audio sources volume.

    - **Invert Polarity** – This filter can be used to fix issues with phase cancellations.

    - **Limiter** – A limiter is a special compressor with an increased attack speed which can be used to prevent an audio signal from peaking. A limiter should be used as the last filter in your chain.

    - **Noise Gate** – A noise gate can be used to reduce background noise based on a level of room noise you wish to remove.

    - **Noise Suppression** – Noise suppression can be used to mitigate room noise from sources such as a computer fan or white noise.

    - **VST 2 Plugins** – Virtual Studio Technology plugins can be used to increase the functionality of OBS. Check out our chapter on using these plugins with OBS.

- **Video Delay (Async)** – This filter will allow you to create a playback delay from your live video sources.

Tip: When you are applying multiple audio filters professionals suggest this audio filter chain. Aligning your filters in this chain will help your voice be more intelligible and make your audio sound more pleasant. Try it for yourself and see!

1. Noise Suppression
2. Gate
3. EQ
4. Compression

- ○ Effect Filters
  - **Apply LUT** – This filter allows you to apply a stylist LUT to any source. A LUT stands for "Look Up Table" and they are similar to an "Instagram" filter which changes the coloring of a video source.
  - **Chroma Key & Color Key** – Chroma Key & Color Key filters are used to remove a certain color from a source. This is mainly used for green screens and each filter performs slightly differently. See our chapter on green screens and virtual sets.
  - **Color Correction** – This filter allows you to adjust contrast, brightness, gamma, saturation

and hue for any source. This filter also allows you to adjust the sources opacity.

- **Crop/Pad** – This filter allows you to crop any source.

- **Image Mask / Blend** – This filter allows you to use an image with a color or alpha channel to mask or blend your source. You can use this to create a circle outline for a live video source like a webcam.

- **Scaling / Aspect Ratio** – This allows you to scale your sources and change the aspect ratio.

- **Scroll** – This filter is generally used to scroll text but it can also scroll any source.

- **Sharpen** – This allows you to sharpen your sources. Try this on a live camera feed to improve you image quality.

When OBS introduces new features, it's always worth checking out the latest filters which allow broadcasters new flexibility all the time. As of OBS 23, even entire scenes can have a selection of filters applied to them. Experiment with the filters that you have in OBS and throughout this book, you will learn some of the most effective filters for video production.

Section 2, Lecture 7: How to Sync Up Your Audio and Video in OBS

In this tutorial, you will learn how to use the free OBS audio and video sync tool included in the online course. Properly syncing up audio and video sources inside OBS is one of the most common problems that people who use OBS have. This tool will help you figure out how much latency there is between your video and audio sources. With this tool you will be able to determine how much delay you need to add to your audio sources to make them sync perfectly with the video.

The tool is a video file which includes audio clicks used for measuring latency recorded via OBS. To get started, you need to play the ten-second video and record it through the camera and microphone you are testing in OBS. Simply point the camera you normally use for your recordings toward the screen where the video is playing and also place your microphone where it can pick up the sound from the video.

The audio/video syncing tool is broken up into three parts. The middle section has a vertical scale that measures milliseconds for accuracy. Beside it is a marker that moves from the top of the scale and falls to the zero mark every second. Every time the marker reaches the zero mark on the scale, the clicking sound happens. On the left of the scale, there is a bigger version of the marker that allows you to read the position of the marker on the scale.

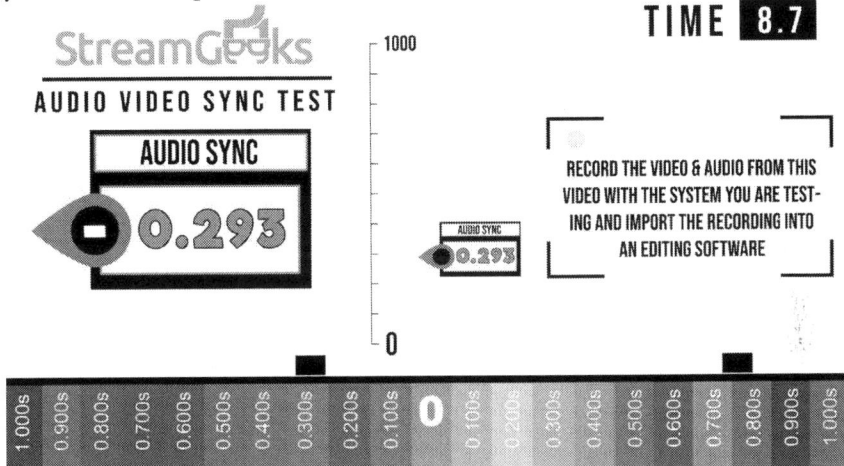

At the bottom of the screen, there is a row of color-coded boxes that gives you an idea of how far your audio source is from zero. You can use this to visually gauge how much delay you need to add to your audio. Usually, audio is processed faster than video inside your computer. That is because audio demands fewer resources from your computer, and the system can process it faster than it does the video stream. As a result, there may be a mismatch in timing between audio and video. The video will play a clicking sound with every one second that passes, and your microphone should capture that sound. You can use the sound recorded in the test video to sync up your audio and video properly. After the recording has been saved, import the video into a video-editing software.

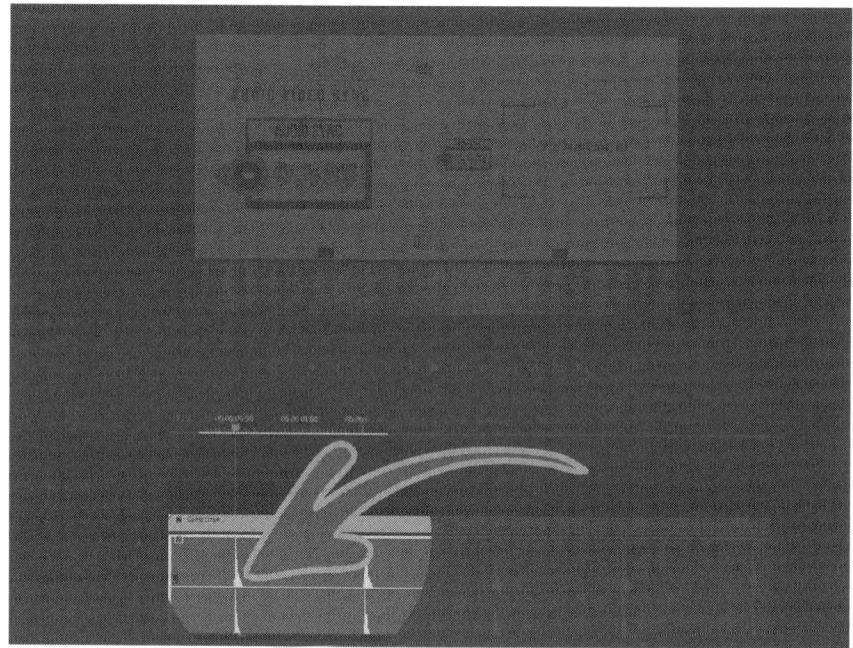

Inside your video-editing software, look at the audio track from your recording. You should notice several peaks in the sound. All those sound peaks occurred every one second in real time. You can use this information to match those peaks in the soundtrack with the video. The idea is to mark the difference between when the sound peaks occur and when the marker on your scale reaches zero on the video. If they do not match, then you must add delay.

To add delay in OBS, go into your audio source, and navigate to "**advanced audio properties**." Look for "**sync offset**" and based on what you see in the video editing software, enter a number to indicate the amount of latency you want in the audio. In the example, about 100 millisecond of latency was added to sync up the audio and video sources.

Pro Video Tip

Don't forget to upload your video content natively to each social media platform. Video always performs better when it's native to the platform (not just a link to YouTube).

Section 2, Lecture 8: Recording Video In OBS

Recording video with OBS is straightforward, and it has a few key advantages you may find useful. If you can capture the essence of your presentation live, it may reduce the amount of post-production you need to do. Most recordings are done either with a producer doing all the switching or a solo operator, using a screen in front of them, while talking to the camera and doing the switching themselves. Depending on your application, having a dedicated producer has obvious benefits.

Let's suppose that you have an OBS setup that includes one camera and a microphone as your main shot. You also have a secondary camera for getting the side angle and a ceiling camera or a desktop capture source to show a document you are discussing. You can quickly switch between these sources using scenes inside your OBS production. If you have a producer working OBS, then Studio mode is ideal because it provides a preview and output for managing up-coming transitions. If you are producing the production yourself, you may want to use the basic mode which will automatically switch the scene when clicked. You can also use hotkeys to trigger videos.

A straightforward way to increase the quality of your video production is to prepare "B-Roll." B-roll is video content that you already made and want to play during your production. B-Roll could be a video with a voiceover that lets you "take a breather" during your presentation. Your B-Roll could also be a video without audio which you play as you read from a script in the background. This gives you the chance to keep the audience from knowing that you are reading from a script while providing visually appealing information you otherwise could not produce live.

To configure your OBS setting for video recordings, go to "**File**" on the menu bar, click on "**Settings**" and then go to "**Output**." Select a file path for where you want your recorded videos to be saved to. When you are recording to your local hard drive, you are not restricted by bandwidth speeds the way you are with live streaming. Therefore, you can choose a recording quality that is better than your stream. Most CDNs such as Facebook and YouTube will already record your live stream in the cloud. So, if you find that your computer cannot handle live streaming and recording at the same time, you can leverage the cloud to store your recording for you. If you are using OBS for recording a video, than you may want to choose a bit rate and format that is of higher quality than your normal streaming setup. For example, our team will generally record our videos in 8-12 Mbps in the mp4 format because it provides a good tradeoff for quality and video storage size.

To configure the advanced settings for your video recordings, you will need to enter the advanced output mode. In the "**Advanced**" output mode, you will see a tab for "**Recording.**" Here you will notice new options for "**Audio**" settings and a "**Replay Buffer.**" You will notice that your live stream can only have one track of audio selected but your recording can have multiple tracks. You can manage the audio sources included in each track inside the main audio mixer on the main OBS dashboard. To open the "**Advanced Audio Properties**" click the cog on any audio source or use the "**Edit**" dropdown menu at the top of OBS. Later, in this course, you will learn more about the Advanced Audio Properties section of OBS.

To start a recording click the "**Start Recording**" button at the bottom right hand side of OBS. You will notice that the CPU usage indicator at the bottom of the OBS dashboard will start to increase. It's worth performing a test recording to see how much CPU usage a video recording with your selected settings demands. This is something you need to keep your eye on when recording. Depending on how powerful the computer you are using is, you might not be able to stream and record at the same time.

Section 2, Lecture 9: How to Add a Webcam and Set Up Picture-in-Picture

Now let's learn how to set up a webcam inside OBS, as well as a picture-in-picture scene. The example here uses a HuddleCamHD Go webcam plugged into the computer's USB 2.0 port. The first step here is to open up OBS and create a new scene. You can rename your new scene as "Webcam Picture in Picture." Then under the sources for that scene, click the plus sign to add a new input and choose the **"Video Capture Device"** as your source type.

Add a Webcam into OBS

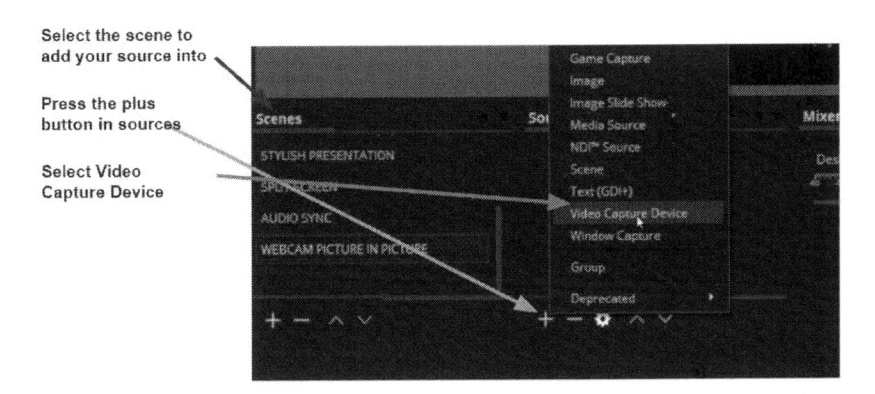

Select the scene to add your source into

Press the plus button in sources

Select Video Capture Device

After you enter the name for your webcam source and click **"Okay"**, and OBS will open a new window with access to all video sources your computer has available. Under **"Device"**, you can click on the dropdown menu to reveal the USB webcam that you want to select. You should know your hardware and choose a resolution, frame rate, and color space your webcam supports. Many webcams allow you to adjust the video feed by clicking on **"Configure Video."**

Configure Webcam into OBS

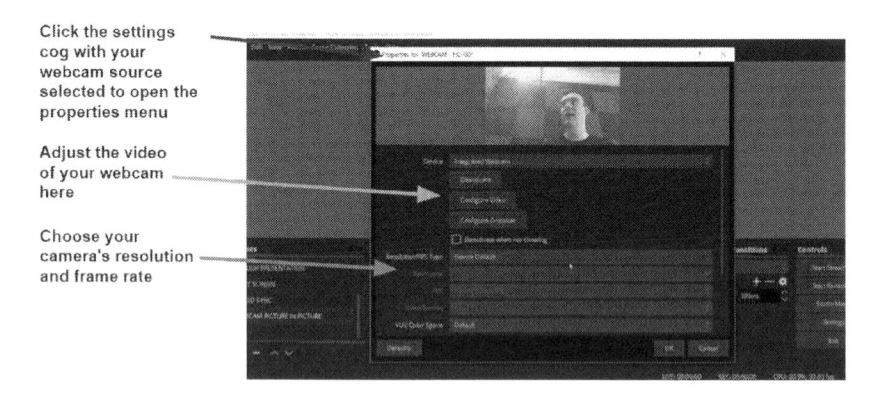

Click the settings cog with your webcam source selected to open the properties menu

Adjust the video of your webcam here

Choose your camera's resolution and frame rate

Next, decide whether you would like to use the "**Device Default**" settings or "**Custom**" settings for resolution and frame rate. Most users will set the frames per second (FPS) to 30 (or 25 for PAL.) Once you add a webcam to OBS you can always adjust an existing camera setting by double clicking the source. In the tutorial example, we also demonstrate how to add an NDI source to your scene with an existing ceiling camera. Since this is a picture-in-picture scene, you can move your camera source up and down in the layering system to choose which one you want on top. You can do this by selecting the camera under sources and using the up and down arrows at the bottom to move it.

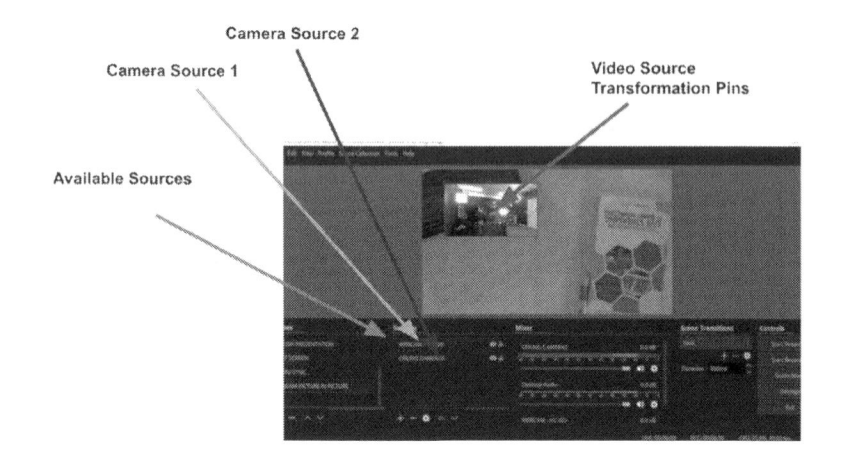

Camera Source 2

Camera Source 1

Video Source Transformation Pins

Available Sources

By now the webcam should be showing on the screen. You may not see the second webcam because the main one will hide it. You will notice an outline at the edge of the webcam picture which can be used to resize the source inside a section of the screen quickly. That will allow your second webcam to be set up in a classic picture-in-picture. Once you have the video feeds precisely where you want, click the lock button on each source to make sure that you do not move them accidentally and mess up your setup.

Section 2, Lecture 10: How to Rotate and Crop Video Inputs

To rotate the camera, first, locate and click it in the sources area. Remember that there are different sources tied to each of your scenes, so you can find the source by identifying the scene that it is connected to. If your source is locked, unlock the source by clicking on the small lock icon. Then you can right click on the camera source and select "**Transform**" in the menu list that appears. A list of options will be shown to you. Select "**Rotate 180 degrees**" and the camera input will change to what it should be. It is that simple!

Sometimes after you rotate your camera, you may discover that the arrangement is not exactly what you want. This can happen, for example, if you have a picture-in-picture setup. You may notice that one video input is partially obscured by the other. You can use the OBS cropping features to enhance your production setup further. Before you crop a video source, it is essential to know that it will be cropped for every scene where you are using it. If that is not what you want, then you should not crop it. But if that is not going to be a problem, here is how you can crop that video input. Right click on the camera source and select "**Filters**" from the menu that appears. In the filters section that will open, click on the plus button, and select "**Crop**." You can crop out portions from the sides, top, and bottom of the video. Enter the amount of space you want to remove from the left, right, top, or bottom of the video.

Section 2, Lecture 11: How to Set Up Color Correction and LUTs

It's now time to learn about color correction and how to apply a LUT (Look Up Table) in OBS. This is important if you have multiple cameras in your studio since you want the color to match on all your sources. The goal is to achieve a level of uniformity for each video input in your production. If possible, you may want to use the same camera model and lens for each video input in your production. You will also want to have all of the cameras set to the same iris and shutter speed settings. Once your cameras have been tuned, you can use the basic color correction tools in OBS to enhance your production quality.

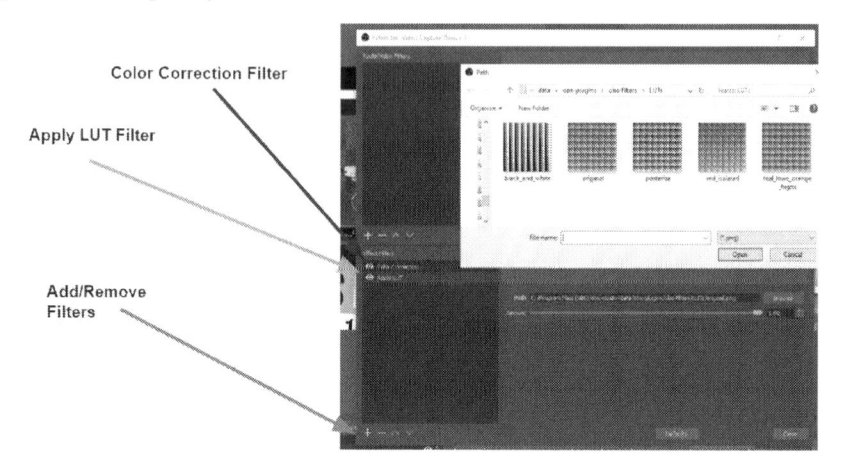

Color Correction Filter

Apply LUT Filter

Add/Remove Filters

Additionally, you can apply identical LUTs to each of your cameras and begin to get them to produce inputs that are even more visually similar. A LUT allows you to take a color scheme from an image and use it to add a filter on top of your video. You can locate the color correction area of OBS by right clicking your source and selecting "**Filters.**" In the resulting window, click on the plus button and select "**Color Correction.**" This will open the OBS color correction options, which include gamma, contrast, brightness, saturation, hue shift, opacity, and color. You can slide the color correction parameters around and use your eyes to gauge how natural the picture looks.

Color Correction in OBS

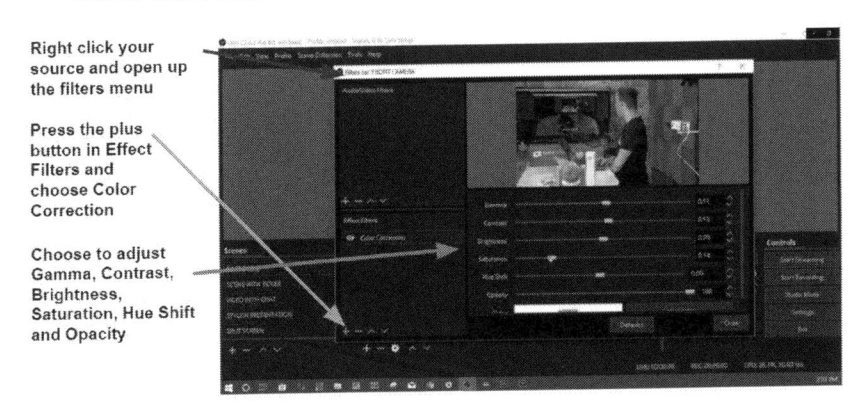

Right click your source and open up the filters menu

Press the plus button in Effect Filters and choose Color Correction

Choose to adjust Gamma, Contrast, Brightness, Saturation, Hue Shift and Opacity

Tip: If you are color matching multiple cameras, you can copy and paste your filter settings from one input to the next. You can do this by right clicking your source and selecting "copy filters." Once you have copied one source's filters, you can paste them into another source.

OBS comes with a selection of free LUTs you can use to apply new filters on top of your inputs. You can find them in the "**Filters**" section by clicking on the plus icon to reveal the "**Apply LUT**" option. You will see a button that says "**Browse**" displayed under the video. Click on it to be taken to the Look-Up Tables inside the OBS installation folders. Click on the one that says "Original" and drag around on the slider to see how the video changes. To get an idea of what it is doing, you can apply another LUT, click browse again, but this time choose the one called "teal-lows;_orange-highs." You will immediately see how much difference the video looks from the other one.

A quick tip to consider when tuning your camera is to adhere to the 180-degree shutter-speed rule. This rule says that your camera's shutter speed should be double the frame rate. For example, if your project is setup for 30fps, then your camera's shutter-speed should be roughly one over sixty. This rule helps to maintain a natural looking video stream from your camera.

Section 2, Lecture 12: How to Use Hotkeys In OBS

Hotkeys are a powerful feature used to perform all kinds of tasks in OBS from your keyboard or USB connected device. OBS includes a long list of functions that can be triggered by your keyboard. As your OBS production becomes more advanced, you will see that assigning hotkeys to everyday tasks will improve your overall video-production workflow.

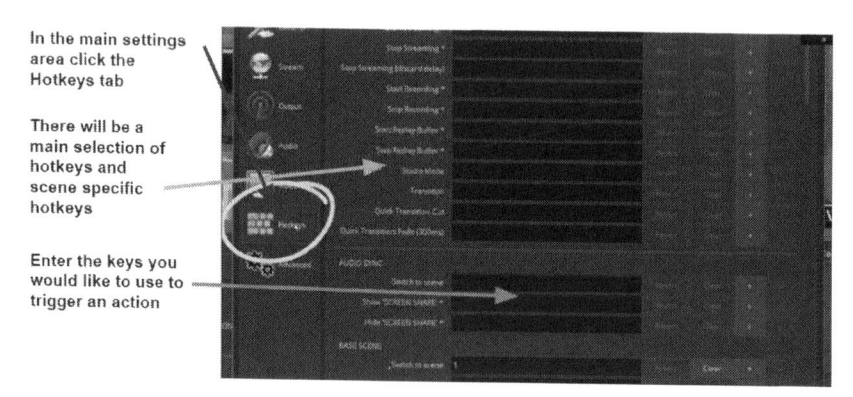

Hotkeys in OBS

In the main settings area click the Hotkeys tab

There will be a main selection of hotkeys and scene specific hotkeys

Enter the keys you would like to use to trigger an action

Hotkeys are managed from the main "**Settings**" areas under "**Hotkeys**." The default list of hotkey function options is quite long, but do not let that intimidate you. Some of the options include start streaming, stop streaming, start recording, stop recording, and transition. Additionally, there are options to create hotkeys for individual scenes and sources inside your setup. If you have multiple scenes that you are working with, OBS will make them available in the hotkeys area as soon as you create them. You can then decide how you want to control the sources for each scene by simply assigning a number or letter to it. For instance, you have a ceiling camera in one of your scenes, you can assign the letter 'Z' to that camera, with a function to show or hide it whenever you want by just hitting the 'Z' key on our keyboard. Consider using your ability to trigger a hotkey with multiple buttons to avoid triggering a hotkey by mistake. For example, instead of using "Z," use "Shift + Z."

Hotkeys can be used to set up transitions. A quick and easy hotkey for transitions is the spacebar. To make the spacebar your transitions hotkey, press the spacebar inside the box beside the transition option on the list. Farther down on that list, under sources, there is also an option for setting up the shortcut for any source. This may be ideal for an intro video. Hotkeys become incredibly valuable if you do not have someone else working with you as a producer. Having hotkeys set up makes your work a lot easier to perform—especially if you are on camera.

Hotkeys are particularly important if you are producing videos for different purposes. PTZOptics, the camera manufacturer, has produced a plugin that allows users to control the pan, tilt, and zoom operations of a camera connected to OBS. Using hotkeys, you can quickly move from one camera to another, without any need to look from your camera to the mouse. Using the PTZOptics camera plugin, you can quickly switch between pan, tilt, and zoom camera presets all within OBS. Many users will set up hotkeys in a way where they only need to hit number one, two, or three to switch scenes. Once this is set up, you can press 0 to play an intro video and space bar to transition between new inputs.

The goal here is to improve productivity, as well as create videos that look highly professional. Hotkeys allow you to switch seamlessly between cameras, scenes, and all the other elements inside your OBS setup. New plugins—like the PTZOptics IP camera control—are starting to extend what is possible with OBS hotkeys.

Section 2, Lecture 13: Creating A Ticker in OBS

Tickers are a great way to show information at the bottom of your screen. Sometimes when you watch the news, you will notice a line of news info that scrolls across the bottom of the TV screen. This information is usually below the newscaster, and it can be set up in OBS to provide your audience with additional information about your presentation.

Setting up a Ticker in OBS

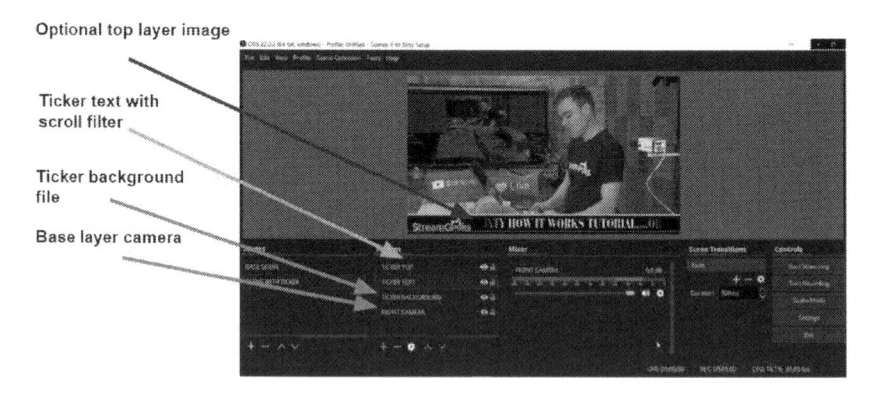

Optional top layer image

Ticker text with scroll filter

Ticker background file

Base layer camera

    The online Udemy course includes some sample files you can use to set up a stylish ticker quickly. Start by adding a black bar to the bottom of the screen that you can display your text on top of. If needed, drag around on the black bar image until it fits at the bottom of the screen. The idea is to create a black background that contrasts with the rest of the screen. Next, add text to the scene. Click on the plus button to add another source and select "**Text**." Rename this to "ticker text" and enter the text you want to display. The text should be sufficiently long, to scroll across the screen. If your message is short, consider adding a break in the text and duplicating your message multiple times in the text field. If you would like to choose a different font from the default font, you may. You can also adjust the size of the text. When you are satisfied, click, "**Okay**."

    Position the text over the black bar image. Now right click on the text in the sources box and choose "**Filters**" from the list of options. Inside the filter box, click the plus sign at the bottom and select "**Scroll**." Now, drag on the slider to choose how fast you want the text to scroll. Click "**Okay**" to close. Your ticker is now ready, and it should be scrolling across the screen. It's that simple!

Section 2, Lecture 14: Using a Green Screen In OBS

Using a green screen with Open Broadcaster Software is a great way to spice up your video production content. A green screen is a green background which can be used with a chroma key filter. A chroma key is a technique used to remove a specific color of any video source so that you can add another background. To do this in OBS, you can first create a new scene and name it 'Green Screen'. You should add a source to the scene. For your camera use the **'Video Capture Device'** source and rename it.

Now right click on your video source and select **"Filters."** Inside the Filters menu, select **"Chromakey."** When you click okay, a set of controls will open with sliders on them. In the first control slider, make sure that 'Key Color Type' is set to green. Start by selecting the color of your screen. You may want to take a screenshot of the color your video background is producing and importing that color into an editing software like Photoshop. Inside Photoshop, you can determine the unique color code your screen is producing. Using the unique color code for your chroma key will perform much better than standard color options for "blue" or "green."

The next level of control is called "**Similarity**." Move Similarity around on the slider to adjust the chroma key. After this, the second control is called "**Smoothness**" and it is used to feather out that faint green outline you may see. Keep adjusting this until it fades away completely, but without making the image look unnatural. Now use the third control labeled "**Spill Reduction**" to remove any bit of green that might be showing on the person in the video. After you have these three controls set as you want them, you may play around with the contrast, brightness, and saturation controls to fine-tune the image to your taste.

One of the most essential parts of chroma key technology is having good lighting. The green or blue background of your subject needs to have an even color for the chromakey to work. If your video shows signs of pixelation, consider using a darker virtual background to mask the noise. The darker the background, the less pixelation you are going to see on the final video.

Section 2, Lecture 15: Making A Virtual Set In OBS

So, what is a virtual set? A virtual set generally extends the idea of a standard virtual background by combining background and foreground elements. It is a sort of simulated reality that you create in your video by inserting an environment that is different from the one in which you are shooting the video. For example, you may be doing your video presentation from your kitchen table but with the help of a virtual set appear to be speaking from inside a newsroom, classroom, courtroom, or even the White House lawn. This simple technique can open a world of possibilities for your productions.

Green Screen Set vs Real Studio Set?

Virtual sets give you the ability to quickly and cheaply customize your setup or choose themes for your videos. If you had to build or travel to every scene you wanted to use, you would be spending a ton of time and money. But with virtual sets, the only limits are your imagination. Wirecast, vMix, and other paid-for streaming solutions include a bundle built-in of virtual sets for their users to enjoy. Because OBS doesn't include any virtual sets by default, most people believe that you cannot make them in OBS, when, in fact, you can. You can find more than 10 free virtual sets to download in our online course. You will also learn how to create your custom OBS virtual sets as well.

Start by creating a new scene in OBS and call it "Green Screen Set," and then add sources to it. The first source will be a video camera. Alternatively, if you do not have a video camera set up, you can use the images included in the virtual set folder provided with this course. To find it, go the hamburger icon on this video or any other video on the course; click it to reveal the resources and content section of the course. Go down the list and you will find a folder called "OBS Virtual Set.zip."

Inside the downloaded Zip folder, are your virtual set image assets. Select the image called "Talent." You can use this as your presenter in the virtual set or use a live camera feed. Also included in the zip folder is a foreground image (a desk) and a background image (a large room). Together these two png files will make up our virtual set. Next, click sources again and select image. Choose the desk image and insert it as the foreground in front of the presenter on the screen. It should make the person look like they are sitting at the desk. Then, drag it around until it fits perfectly in front of your presenter in the video or image, whichever you are using. Next add in the background, which in this case is a classroom background. Go to sources again, choose **"image"** and select the background image from the zip folder that you downloaded. Adjust it until it is the appropriate size.

You may have to reduce the size of the desk to have it fit into the room, until it looks like it is sitting inside it. Then transform the video or image of your presenter to a size that is proportionate to the dimensions of the table. There are several sites that offer free virtual sets that you can use in OBS; just search around the internet. Download some of those and play around until you get the hang of making your own virtual sets.

Section 2, Lecture 16: How to Use the OBS Multiviewer

The new OBS "**Multiviewer**" is an incredibly powerful tool for viewing multiple scenes inside your OBS production simultaneously. The Multiviewer is an essential tool for OBS producers and camera operators. For a long time, the Multiview feature was only available in premium paid for software like vMix or Wirecast. But with Multiview now built into OBS, you can set up a secondary monitor or window to view your top eight scenes with preview and output windows.

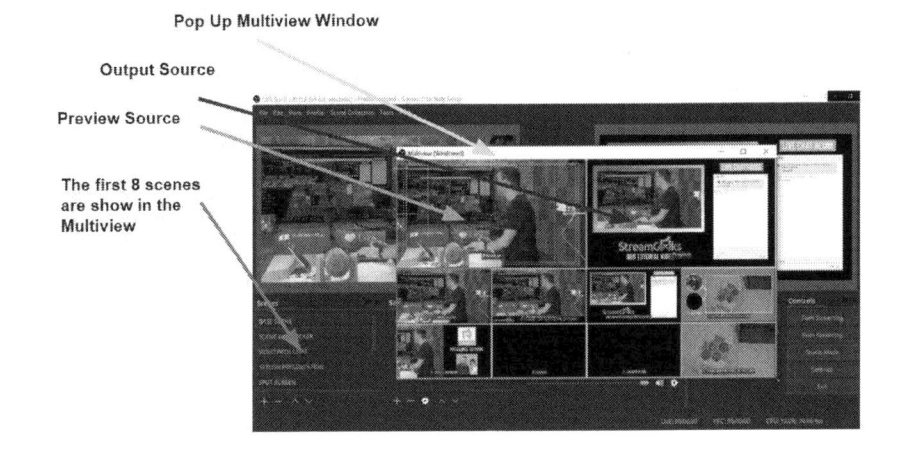

In the main OBS dashboard, click on the "**View**" tab. Inside the view tab, you will find two options for Multiview. You can choose from "**Multiview (Fullscreen)**" and "**Multiview (Windowed)**." If you click on the windowed options, all of your sources will immediately become visible in a new pop up window that appears on your display. You will be able to view up to eight scenes within that one window. You can also move this window around to position it anywhere on the screen. In the "**General**" settings area you can set up OBS to "**Click to switch between scenes.**" This will allow you to click on any of the eight scenes in the Multiview to quickly transition to that scene. If you have hotkeys enabled, you can easily switch between sources while you monitor what you are doing in the multiview window that you have open. The multiview window can also be manipulated to suit your preferences. You can place it anywhere on the screen, stretch it to whatever size you need it to be and even have it visible from a second or third monitor. If you are a producer, camera operator, or working with a team, having this functionality available makes your work a lot easier. This feature is especially helpful if you are working with multiple monitors.

You can also use the Multiview feature in full screen mode. When selected in full screen mode the Multiview display will take over an entire screen. You will see that the Multiview full screen option will give you the ability to choose which monitor you would like to use in full screen mode. With Multiview in full-screen mode, you receive more details of what is happening in every scene and you can make a better decision of which camera to show to your audience. You will notice most high-end video productions use Multiview windows in full-screen mode.

Section 2, Lecture 17: How to Create and Use Stingers In OBS

Pro Video Tip

Most professionals will use simple cuts between live camera shots. Using too many fancy transitions can actually make your production look amature.

OBS now supports stinger transitions. Stinger transitions are animated video files that are used to create fancy transitions between scenes. The effect combines a transparent video animation that evolves into a full screen overlay which is timed with a transition. When your stinger video animation starts playing on top of your current scene, you can have OBS cut to your preview source exactly when the video completely covers up the current scene.

The stinger effect has been made popular by sports broadcasters who use the effect to notify audiences of specific scene changes. In this way, sports broadcasters have trained their audience to expect a stinger transition when an instant replay or prepared video screen of statistics is coming up. Fancy stinger video animations often include perfectly timed audio which include lots of "whooshes" and "pops" to give the animation a realistic feel. Remember that most viewers have never heard of a "stinger effect," yet the nonverbal connection that is made to the audience takes effect immediately.

You can create a custom stinger transition inside Adobe After Effects and import your work into OBS for testing. If you do not have Adobe After Effects, you can consider skipping this portion of the tutorial. It may be easier to hire a freelancer on a website like Fiverr to create a custom animation for $5-10.

To build a stinger animation with Adobe After Effects, you want to create a 1920x1080 scene with an object that slides into the scene quickly, makes a brief appearance, and then covers the entire area on exit. At the moment this stinger takes up the whole screen, OBS will initiate a cut to the scene in preview. In this way, the stinger can create an animation that displays itself over any current scene you want to use, and transition after you animation is complete.

Go ahead and open Adobe After Effects to create a new 1920x1080 composition that is two seconds long. Next, you can drag and drop in a .png file of your logo into After Effects to be used as your main file to animate. Drag and drop this logo into your composition and click the drop-down arrow next to the title line to reveal the "**Transform**" area. The transform area of your png logo file is where you will tell After Effects how you want your logo to move throughout the two seconds of the composition. Choose the "**Scale**" transformation option. To start a transform session, click the clock which will view your animation over time and create a keyframe for you to use at the very first frame of your composition. Next, you can drag the timeline slider over a few tenths of a second. Let's scale your image to fill up the whole screen and make a new keyframe there. You can scale the logo down to a smaller size and After Effects will animate the rotation in between each of these two keyframes.

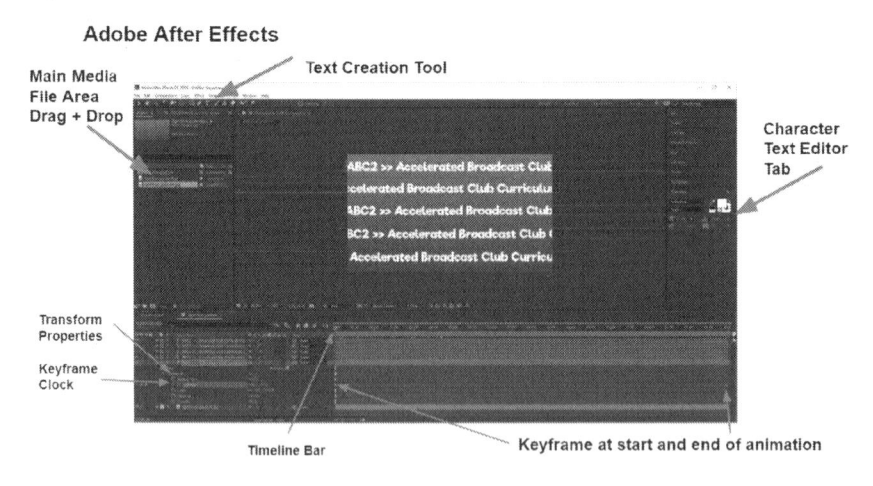

Adobe After Effects

Main Media File Area Drag + Drop

Text Creation Tool

Character Text Editor Tab

Transform Properties

Keyframe Clock

Timeline Bar

Keyframe at start and end of animation

Now, let's add a rotation into the mix. You can create a transformation keyframe at the zero mark and another keyframe exactly where your logo stops scaling. Finally, you can end the composition by scaling the logo past full screen at the end of the two-second composition. This is the point where the logo will cover the entire screen and perform the cut to your next video input in the preview. Preview the animation inside After Effects to make sure everything is animating as planned. You can move the keyframes around in the timeline to tweak the movements of the animation. Now that the animation is complete, you can export this video as an avi file. Make sure to click the **"RGB + Alpha"** channel option when you are outputting a transparent video file from After Effects.

To set up your stinger video transition inside OBS, first, make sure you have the most up-to-date version of OBS (23 or later is required). Then click on the plus button inside the scene transitions area and select **"Stinger."** This will open up a properties box where you can choose your avi file and choose a transition point for your stinger. Since the stinger is two seconds long, choose 1900 milliseconds for your transition point, which is just one-tenth of a second before the animation ends. The transition point is the time that OBS uses to cut between the preview and output scenes regardless of how much time is left on your animation. In this way, you want the animation to play into the next scene. You can now click the plus button in between the preview and output screens in studio mode to add the stinger to your button list.

TIP: You can set up a hotkey in the settings area of OBS for transitions. Consider using spacebar as a hotkey for quickly cutting between two scenes. Once this is done, you can set up another hotkey just for the stinger effect.

That is how to make a stinger with your custom logo on it. However, if you do not want to bother with Adobe After Effects, you may go to videoblocks.com and get a stinger there.

Section 2, Lecture 18: How to Enhance Your Audio with VST2 Plugins and Voice Filters

A big part of your live stream is your audio, and many streamers struggle with producing good quality audio. Without good quality audio, all the work you put into creating a nice-looking presentation may be in vain. If the audio in your production is of poor quality, your audience will lose interest. A common beginner mistake is to have sound coming in from various sources with levels that do not match. Luckily, OBS supports Virtual Studio Technology 2 (VST2), which can be used to help improve your audio considerably. You can download a set of free beginner plugins at https://www.reaper.fm/reaplugs/. This set of free plugins comes with a compressor, noise gate, delay, advanced equalizer, and an array of other tools.

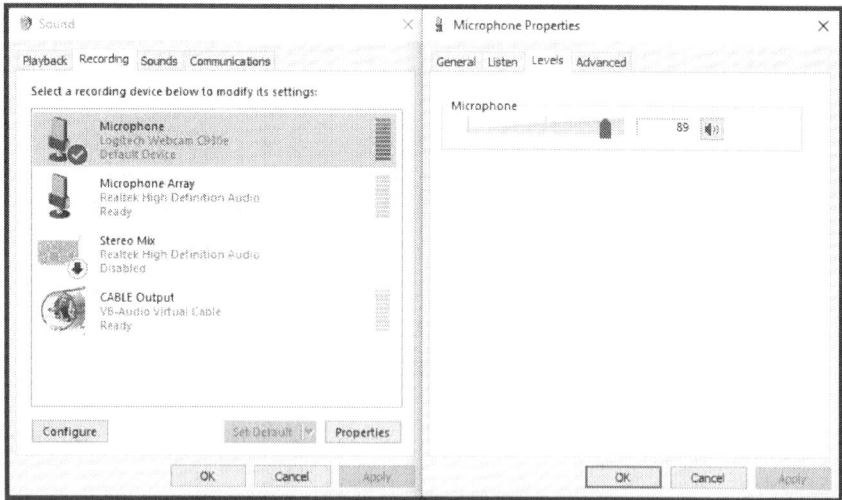

An excellent place to start with audio when it comes to OBS is your computer's default "**Sound Control Panel.**" In Windows, the sound control panel has a tab for "**Recording**" which reveals all the microphones connected to your system. You can tweak the default settings for your microphones here. Use the active green sound level indicators to set the baseline levels in your system. Keep an eye on the green indicators as you adjust and optionally boost levels. As you are doing this, make sure that the level on the indicator never peaks.

Once the microphone levels coming into your operating system have been set up correctly, you can go into OBS and make one of your microphones the default audio source. In the OBS dashboard, go to "**File->Settings->Audio,**" and you will be able to see the microphone sources you have connected. Choose the audio sources you want as your defaults. Next, you can start to add filters to your audio sources to enhance the quality inside OBS.

OBS comes with basic noise-suppression tools that are easy to use. You will find that the VST 2 noise gates from Reaper are considerably more advanced. Click on the cog on your mic under the mixer section of the dashboard, select **"Filter,"** and choose VST 2 Plugin in the menu. With your plugin ready, you can proceed with setting up the noise gate. Noise gates are great because they reduce background noise that is audible under a certain gated level. To determine the gated level, listen to your microphone input when the room is totally quiet. You can watch the levels on the indicator; without saying anything at all, and you will identify the "room noise" level. This noise level in the room tells you where to set the gate. The gate will mute the microphone at that point so that if you are not talking, there is no distracting room noise.

The next useful plugin used to enhance your audio quality is the Equalizer or EQ. Once again, select **"Filters"** and choose the VST 2 EQ plugin. Now select **"Reaeq Standalone"** and click on **"Open Plugin Interface."** On the visualizer that opens, you will see the audio levels going up and down in response to your voice or whatever sound is present in the room. Using the EQ, you want to enhance your voice by removing portions of the audio that you do not need and leaving it in the range of what you use. An EQ helps remove low-hum sounds from passing busses or high-pitched sounds from electronics. These noises that are picked up by your microphone do not generally fall in the range of the human voice.

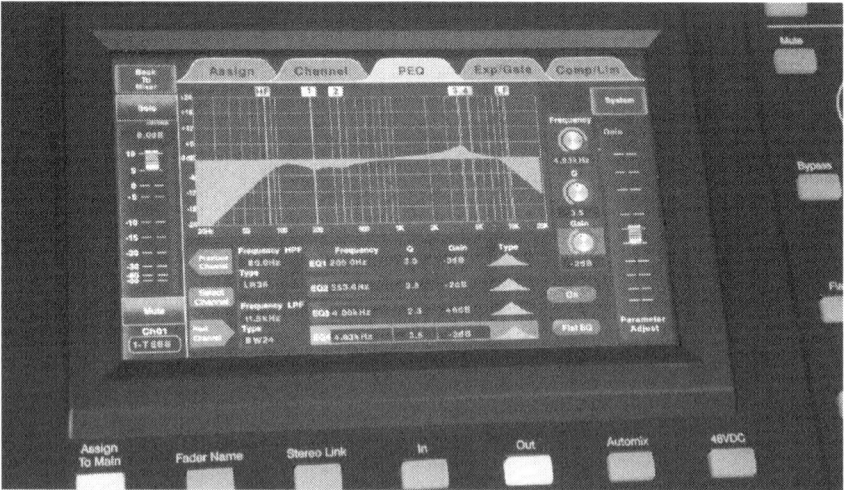

A basic EQ generally starts with a low shelf and a high shelf. The human voice has a range of 80 Hz- 8,000 Hz and this is usually the most important area of an EQ. Therefore, a low shelf can be setup to remove all noise below let's say 70Hz. A high shelf can be set up to remove anything above let's say 10,000Hz. To get the best possible sound, monitor both the equalizer and the microphone level indicator. To properly set EQ and audio input, it takes a lot of practice and actual listening. Active listening is a skill that audio engineers develop over many years. Using basic low-shelf and high-shelf EQ tools is a significant first step toward improving your audio quality. Remember that every voice is unique. It might take some time, but the rewards that you get in sound quality are well worth the effort. Again, even though, at first glance, it looks intimidating, it is not that difficult to make incremental improvements to your audio.

Section 2, Lecture 19: How to Use RTSP Streaming in OBS + PTZ Camera Controls

In this tutorial, you will learn how to use RTSP IP video streams in OBS. RTSP video streams are ideal for streaming video over your LAN (Local Area Network). RTSP video does not require a traditional HDMI or SDI to USB capture card and it allows you to plug your camera directly into ethernet to bring it into OBS. For this tutorial, you will expand your knowledge of plugins using the PTZOptics camera controller plugin for OBS. This plugin allows you to use a single ethernet connection to your camera to receive video in OBS and control the robotic movements of a PTZOptics camera.

There are currently two ways to bring RTSP video streams into OBS. It is possible to bring in RTSP video sources into OBS using the "**Media Source**" input option. However, there have been reported issues with this method, which cause temporary disconnection with RTSP video randomly. A more reliable way to connect RTSP video feeds into your OBS production uses a free software called VLC Media Player. Once you install VLC Media Player, you will automatically see "VLC Video Source" as an option in your OBS sources menu. Once your VLC Media Input is open, click the plus button on the right and select "**Add Path/URL.**" Here you can enter the RTSP address for your camera. Generally, this address will look like "rtsp://THE-IP-ADDRESS-OF-YOUR-CAMERA/1" OR "rtsp://192.168.1.99/1." Before clicking "**OK,**" leave "**Loop Playlist**" checked and "**Always play even when not visible**" on. You may find that your cameras have 1-2 seconds of delay. For more reliable and lower-latency IP video connectivity, you may want to look into NDI (Network Device Interface) cameras and sources. Remember that you can always match up your video sources with the incoming audio by adding an audio delay.

Now you can get started with installing your first plugin in OBS. The first step is, of course, to download and install a plugin. For this example, use the PTZOptics OBS plugin available at ptzoptics.com/OBS. Once you have downloaded this plugin, follow the instructions in the read.me file which outline where your need to copy the plugin files into the OBS application folders. Once installed, you can open OBS and go to the top menu on the dashboard and click "**Tools**." You should now have an option that says "**PTZOptics Camera Controller**" listed. Click on it to open the PTZOptics control panel. You can type in the IP address of your camera and hit the "**Connect**" button in the settings menu to allow OBS to control your IP connected camera. Now you can control your camera using the plugin. There are buttons on the panel for issuing various commands such as camera pan, tilt, and zoom. With this tool, it is possible to remotely control as many as eight cameras inside your OBS production. Finally, a new update to the PTZOptics camera control app makes these cameras controllable with a USB XBox joystick!

Section THREE

Section 3, Lecture 20: NDI For OBS Overview

One of the most advanced new features of OBS is a plugin that provides an integration with the NewTek® NDI®. This feature brings the tremendous power of IP video to OBS. IP video is unarguably the future of video production. Working with IP video will open many opportunities for video production that will reduce overall implementation costs and increase what broadcasters can do on a budget.

If you have ever used a Serial Digital Interface (SDI) or RS-232 cable, then you know what it's like to work with video production equipment in the analog days. SDI cabling, which is still used heavily in the broadcast industry, is used to provide substantial bandwidth pipelines for dedicated video equipment. The issue with SDI is that it requires expensive converters to make this video input available to a computer using OBS. Ethernet connectivity is preferred because all computers, these days, are already connected to Ethernet. For most OBS producers, it makes perfect sense to transition to a workflow that includes video over IP once they see the ease of implementation and video quality.

Most of the devices we use—whether they are Mac, PC or Linux computers, wireless tablets, or smartphones—can be all connected to a a local area network (LAN). In most cases, OBS is also connected to the local area network and can therefore quickly gain access to network-connected devices. This means that schools, houses of worship, businesses, and other organizations, which already have extensive local area networks, can use those same networks for video production and distribution.

One of the great things about NDI is its widespread adoption. So many of the best live streaming software platforms already support NDI, including vMix, Wirecast, Livestream Studio, VLC, OBS, and even Skype. So, with NDI in your personal or organizational video production setup, you have access to a world of possibilities for video production and live streaming.

The main requirement OBS broadcasters need to know about regarding NDI is the need for a gigabit network infrastructure. A 10/100 Ethernet setup usually transmits data at 100 Mbps and does not have enough bandwidth to support an IP video setup. With a Gigabit network in place, you can support up to five or six NDI streams and even more NDI HX (High Efficiency) streams on the network. NDI|HX video sources are generally one tenth of the bandwidth of full NDI sources. Review the bandwidth comparison chart below.

NDI Mode	Bandwidth
HDI HX Low (720p60fps)	6 Mbps *

NDI|HX Medium (1080p30fps)	8 Mbps *
NDI|HX High (1080p60fps)	12 Mbps *
NDI (1080p60fps)	125-200 Mbps *

* (Nominal Range). NDI improves with each new release just like OBS. Quality continues to increase with NDI and sometimes bandwidth requirements can vary.

One interesting use case for NDI video outputs from OBS is simulcasting. Using a single OBS encoder, you can only live stream to a single destination. Most users choose either Facebook or YouTube, but not both at once. However, if you have your OBS system set up with an NDI output, you can have a second PC using OBS to stream to a second CDN. To do this, you can set up your second computer to receive the output of your first OBS production. The second computer can then encode and stream this output to any second CDN destination that you would like.

Simulcasting - YouTube and Facebook (4.5Mbps Total)

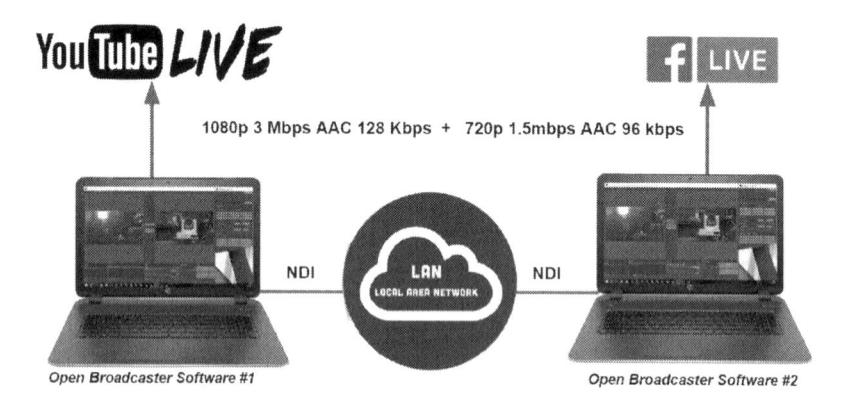

YouTube LIVE   LIVE

1080p 3 Mbps AAC 128 Kbps + 720p 1.5mbps AAC 96 kbps

NDI   LAN   NDI
       LOCAL AREA NETWORK

Open Broadcaster Software #1   *Open Broadcaster Software #2*

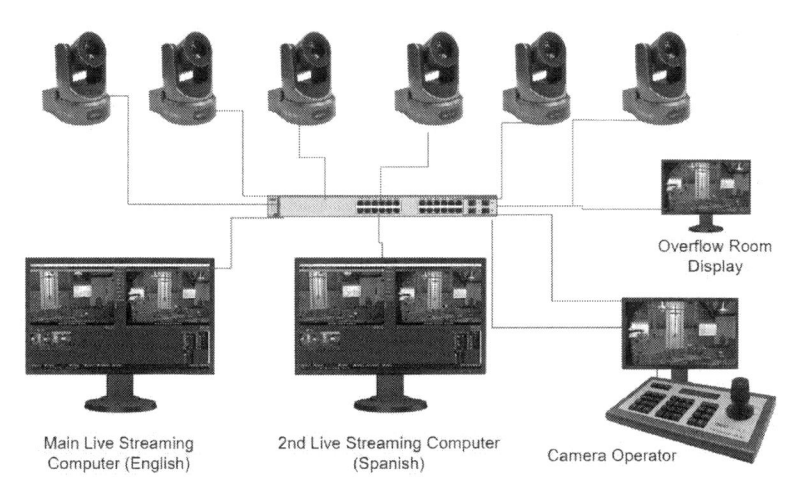

Main Live Streaming Computer (English) 2nd Live Streaming Computer (Spanish) Camera Operator

Overflow Room Display

Let's talk about the basics, to make sure you have a good understanding of what IP networking involves. Ethernet cables connect each device to a network switch which acts as a hub inside your LAN (Local Area Network). A local area network is a group of computers and associated devices that share standard communications lines or wireless links to a server or router. Every device on your network has an IP address which Wikipedia calls "An Internet Protocol address . . . a numerical label assigned to each device." An IP address generally looks like this: 191.168.1.100 but it could also look like this 216.3.128.12.

Look at the following example IP Address Table.

IP Address	Device
192.168.1.0	This is the network number that identifies the network as a whole
192.168.1.1	This is assigned to the router
192.168.1.2-254	These addresses may be assigned to devices on your network
192.168.1.255	This is the broadcast address. Anything sent to this address is automatically broadcast to IP addresses 1-254

As you consider implementing an IP based video production setup, you may want to consult with your information technology department. If the IT department is unsure whether you can leverage the existing network for IP video, you may need to set up a dedicated network. Either way, it's essential to understand how this technology works.

In this example, there are several parts of the network segmented out for devices that are used for video production such as cameras, video production systems, and computers. Every device on the network will have an IP address which will help other devices communicate with one another on the network. There are two different ways that you can assign IP addresses to devices. They can be assigned static IP addresses manually or dynamic IP addresses automatically. Static IP addresses never change, and, therefore, they are much better for managing an IP address table on your network. Dynamic IP addresses are assigned by your router using DHCP (Dynamic Host Configuration Protocol). DHCP is ideal for devices that periodically connect and disconnect from your network. A prime example of an IP connected device that uses DHCP is a smartphone. When your smartphone connects to WiFi, it automatically gets an IP address from the network. It's considered best practice to assign static IP addresses to the most critical devices on your network used for video production.

Without getting too complicated into networking jargon, you can have up to 254 devices on a single network which can all communicate on the same IP range. The network above as a whole is defined as 192.168.1.0 and the router would usually be assigned the very first address as 192.168.1.1. Your router is generally given to you by your ISP (Internet Service Provider), and this device may include a built-in network switch, a firewall, and a Wi-Fi access point. Therefore, many routers today will allow you to connect devices to your network right away. You may have seen a router at home connected to devices such as a smart TV, your smartphone, and perhaps a few computers. For video production, you will likely want to purchase a dedicated network switch which will allow you to plug multiple devices into your network.

Remember that ethernet cables can become a bottleneck in your bandwidth access. Always select ethernet cables that match your networking infrastructures. If you are planning to power cameras using ethernet cables, you will want to make sure to purchase a network switch that supports PoE (Power Over Ethernet).

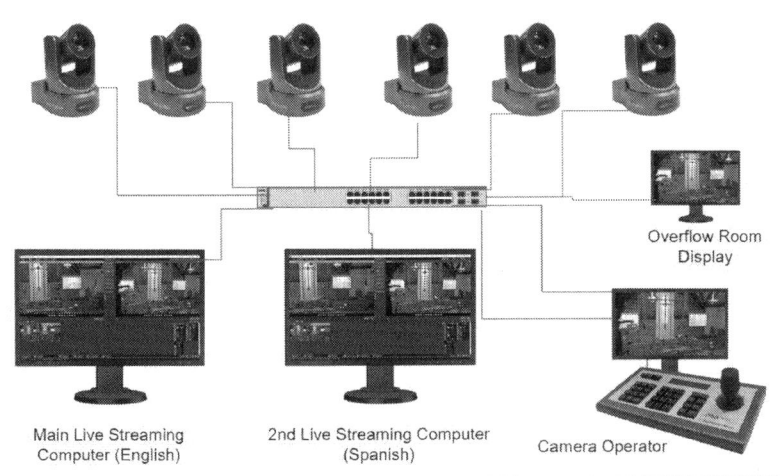

IP Address	Device
192.168.1.0	Network Address
192.168.1.1	Router supplied by your Internet Service Provider
192.168.1.2-59	Used for office devices like printers, access points, and other IP connected devices
192.168.1.60	PTZOptics 20X - Main Camera Studio Camera
192.168.1.61	PTZOptics 12X - Secondary Camera
192.168.1.62	PTZOptics 12X - Main Weatherman Camera
192.168.1.63	PTZOptics 20X - 2nd Camera Weatherman Camera
192.168.1.64	PTZOptics ZCam - Static Camera used for Behind the Scenes
192.168.1.65	PTZOptics ZCam - Static Camera used for live clock image
192.168.1.66	PTZOptics IP Joystick Controller

192.168.1.70	Main Live Streaming Computer
192.168.1.71	Laptop with Weatherman Sliders
192.168.1.72	Computer Powering 2 Displays for confidence monitors
192.168.1.73	Computer Powering a display for camera operator 1
192.168.1.74	Computer Powering a display for camera operator 2
192.168.1.119*	iPad using NDI Camera App (Wireless Camera)
192.168.1.123*	Smartphone used for iOS camera control app
*Assigned with DHCP	

The great thing about IP based video production for so many streamers is that you already have a network in place. This is where the rubber meets the road. Unlike SDI and HDMI cabling, ethernet can provide power for cameras using a PoE source, such as a PoE switch. It will simplify installations and eliminate the need for additional outlets where you would have had to hire an electrician in the past. And unlike traditional RS-223 control cables, ethernet can also be used to control cameras and devices within your favorite video production software. You can even use an IP joystick without requiring direct analog control cabling to each camera like before.

Now for some of the technical knowledge surrounding network infrastructure and setup. It's time to understand the bandwidth requirements on your network infrastructure for IP based video distribution. Just like the category cabling mentioned early, networking equipment has bandwidth limitations. Most commonly installed networking equipment has a bandwidth limit of either 10/100 or gigabit. Unfortunately, if you have 10/100 networking infrastructure, you cannot use it for IP based video production with the NewTek NDI. There isn't enough bandwidth on these older networking systems to support video transmission. The good news is that gigabit networking equipment has become the industry standard, and there is a good chance this is the type of technology you already have installed. A gigabit network switch with a full throughput backplane can send 1,000 megabits of data to each device on your network. You should never use 100% of the available bandwidth on your network because you need to reserve "headroom" to avoid network congestion and failure. Network bandwidth headroom recommendations can vary widely, but generally, most IT professionals recommend 30% - 60% depending on what the network is utilized for. You should consider consulting your network administrator before adding IP video traffic on to your local area network. Newtek suggests, "NDI traffic should not take up more than 75% of the bandwidth of any network link" (NewTek, 2016).

There are many different types of network switches that can support various levels of bandwidth. While gigabit is the most popular, today you can purchase 10-Gigabit ethernet switches that provide 10,000 megabits per second of transfer speeds. As time moves forward, access to higher bandwidth devices will become more and more common. It's incredible how far we have come already, and things are moving faster than ever before.

Now let's take a moment to understand the bandwidth options you have so that you can optimize our network for the video sources you want to use. Since the NewTek NDI as the example IP video production solution, let's look at the two main types of NDI video: NDI and NDI HX. NDI is considered the full bandwidth compression version which can take a three gigabit, fully uncompressed video signal, and compress it down to 125-200 megabits without producing noticeable digital artifacts. This type of compression is what makes IP video production possible on a gigabit networking infrastructure. The compressions effect is "unnoticeable" to the human eye and completely un-noticeable once the video reaches its final destination. Since the destination for much of our live video sources is a content delivery network like Facebook and YouTube, we know that the video is going to be compressed anyway via RTMP before it reaches our end viewers. The compression technology today is so good that the benefits of uncompressed video are only reserved for the highest-end television video production studios and Hollywood producers.

To further advance what is possible with IP based video production, NewTek released a "High Efficiency" version of NDI called "NDI HX." This version of NDI can compress a 1080p video source down to mere 8 - 20 Mbps depending on the selected quality. NDI HX is available in compression ratios of low, medium, and high. Let's look at the differences in bandwidth using the chart below.

NDI Mode	Bandwidth
HDI HX Low (720p60fps)	6 Mbps
NDI|HX Medium (1080p30fps)	8 Mbps
NDI|HX High (1080p60fps)	12 Mbps

NDI (1080p30-60fps)	125 - 200 Mbps (Nominal Range)

As you can see, there is a big difference between using "full NDI" and NDI HX sources on your network. If you plan on using a lot of NDI sources on your gigabit network switch and respecting the recommended 30 - 60% headroom space for reliability, your available bandwidth can quickly get used up. Let's look at an example bandwidth consumption table below.

Example:

NDI Device Examples (1080p60fps)	Bandwidth	Accumulated Bandwidth	Total % of Gigabit Network Switch
NDI Scan Converter on Laptop for PowerPoint slides	125 Mbps	125 Mbps	12.5%
2 x NDI Monitor for camera operators	125 Mbps / Each	375 Mbps	12.5% / Each
OBS System output in 1080p 60fps	125 Mbps	500 Mbps	12.5%
NDI Monitor in Overflow Room	125 Mbps	625 Mbps	12.5%
5 x PTZOptics NDI|HX (High)	12 Mbps / Each	685 Mbps	1.2% / Each
Suggested Headroom	250 Mbps	910 Mbps	25%

Total Usage			91%

Multiple types of IP video streaming are used in video production today. The most common of these IP streaming types are RTSP, RTMP, and NDI. Let's talk a little bit about RTSP and RTMP. RTSP stands for real-time streaming protocol, and it is a widely used protocol for streaming video and audio on your local area network. RTSP is perfect for viewing a live camera that is sitting on your local area network. RTMP stands for real-time messaging protocol and it used by CDNs such as YouTube and Facebook for streaming your live production over the public internet. It's important to think about the differences between IP video that is on your LAN (Local Area Network) and video that is sent over the WAN (Wide Area Network). I have included the following diagram below to help illustrate this process.

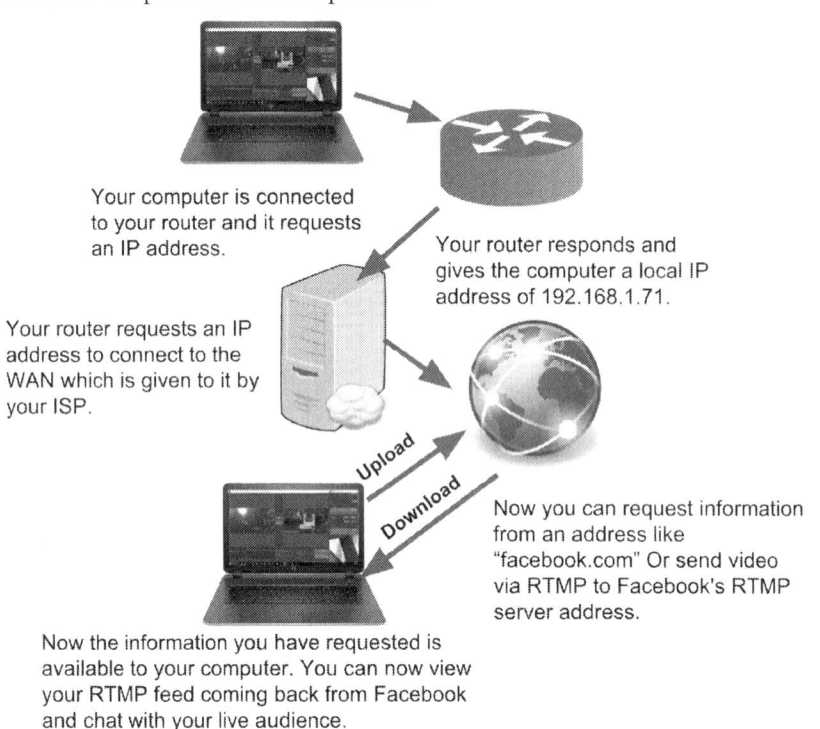

Your computer is connected to your router and it requests an IP address.

Your router responds and gives the computer a local IP address of 192.168.1.71.

Your router requests an IP address to connect to the WAN which is given to it by your ISP.

Upload
Download

Now you can request information from an address like "facebook.com" Or send video via RTMP to Facebook's RTMP server address.

Now the information you have requested is available to your computer. You can now view your RTMP feed coming back from Facebook and chat with your live audience.

Now it's time to talk about multicast network traffic. This is as advanced as we will get in this book when it comes to network traffic. But this is an incredibly important technology to understand when it comes to IP-based video production. Multicast is a method of sending data to multiple computers on your LAN without incurring additional bandwidth for each receiver. Multicast is very different from Unicast, which is a data transport method that opens a unique stream of data between each sender and receiver. Multicast allows you to broadcast video from a single camera or live streaming computer to multiple destinations inside your network without adding additional bandwidth burden on your network for each receiving device. Broadcast studios are using multicast to have a camera operator on one computer viewing the camera feed and the live streaming computer using the same feed at the same time.

Funny enough, this is a technology that many of us have at home built into our television receiver boxes. When you request an on-demand video from your cable television provider, this opens a unicast stream for that unique video. When you are flipping through the hundreds of available television channels, this is using multicast. This is how your cable television provider can send thousands of video channels to your television using the same ethernet cabling available from Best Buy.

MODEL	AREA	POE	IP ADDRESS	MULTICAST ADDRESS
PT20X-NDI-GY	FRONT	Y	192.168.100.31	234.1.0.31
PT20X-SDI-GY-G2	WEST	Y	192.168.100.32	234.1.0.32
PT20X-SDI-WH-G2	EAST	Y	192.168.100.33	234.1.0.33
PT20X-SDI-WH-G2	BEHIND THE SCENES	Y	192.168.100.34	234.1.0.34
PT-JOY-G2	BROADCAST AREA		192.168.100.35	N/A

OPEN BROADCASTER SOFTWARE	BROADCAST AREA	N/A	192.168.100.113	N/A
PT12X-SDI-WH-G2	SIDE	Y	192.168.100.41	234.1.0.41
PT20X-SDI-WH-G2	SIDE	Y	192.168.100.42	234.1.0.42
MAC MINI	FRONT STAGE	N/A	192.168.100.43	N/A
INTEL NUC	LOBBY	N/A	192.168.100.44	N/A
IP Joystick	2^d Floor		192.168.100.51	N/A

    In the diagram above, we can see that there are 6 multicasts enabled PTZOptics cameras. Because these cameras are enabled for multicast, the video feeds can be used simultaneously by multiple computers on our network. Therefore, we can do things like setting up a second computer dedicated to a production made in Spanish. Perhaps, we have a translator in our broadcast club providing Spanish translated audio available to our 2nd live-streaming computer. We can also simultaneously have a camera operator pulling in the video feeds along with an overflowing display in another room.

    The IP address table above also includes information about the device's multicast addresses. Multicast addresses can range from 224.0.0.0 to 239.255.255.255. If you want to leverage the power of multicast video, you must remember to select networking equipment that can support multicast video. Also, remember that each video device that you plan to use will require its own multicast address. Notice that each camera has a unique IP address and a unique multicast address.

    Finally, when you are selecting a network switch to be used for the NewTek NDI, we highly suggest considering a switch that meets all the following requirements. You may want to set up a dedicated network just for your IP video sources especially if this is the first IP-based video production system you have setup ever.

- Gigabit Ethernet *Required*
- Full Throughput Switch Backplane *Required*
- DHCP Recommended
- For Devices That Optionally Support PoE
  - PTZOptics NDI|HX Cams require PoE (15.4w)

  - NewTek Connect Spark requires PoE (15w)

  - Note* PoE+ supports PoE, but PoE doesn't support PoE+

  - Make a note of the power needed for devices/switch

IF YOU ARE USING A MANAGED SWITCH
Managed switches are great, but the settings need to be tweaked to accommodate low-latency IP-based video for production. You can use almost any Gigabit managed switch that meets the requirements above, but you will also have to disable a few settings and enable Flow Control as Asymmetrical.
- Disable Quality of Service
- Disable Jumbo Frames
- Enable Flow Control as Asymmetrical of Simply as On
- Enable IGMP Snooping if Using Multicast (mDNS)
- Configure IGMP Querier and Query Interval Per Switch in Multi-Switch Networks (While Using Multicast)

DEALING WITH FIREWALLS
- mDNS must be accessible
- Manual discover requires access to port 5960 for messaging and all coming after 5961 for streams
- Check the port range from Microsoft PCs using Cmd: ntsh

NETWORK ADAPTERS
- Use DHCP to assign IP addresses or assign static manually
- Use manual configuration in NDI Access Manager to cross subnets
- Designate network location on all NICs as Work (private)
- Connect and make available Gigabit + network interfaces

A little bit on latency
- Full circle latency must be <14ms
- NDI v3.5 supports UDP with forwarding Error Correction for unicast (prior versions use TCP)

If this all looks like another language, please don't worry. I have an entire course on the NewTek NDI linked below. NDI was made to make IP video easy and you may not ever need to play with IP addresses. It's likely that you already have a networking system in place. If you're going to leverage that system as a video distribution network, you need to configure some of the network settings.
https://www.udemy.com/newtek-ndi/

Section 3, Lecture 20: Using NDI Cameras & Studio Monitor with OBS

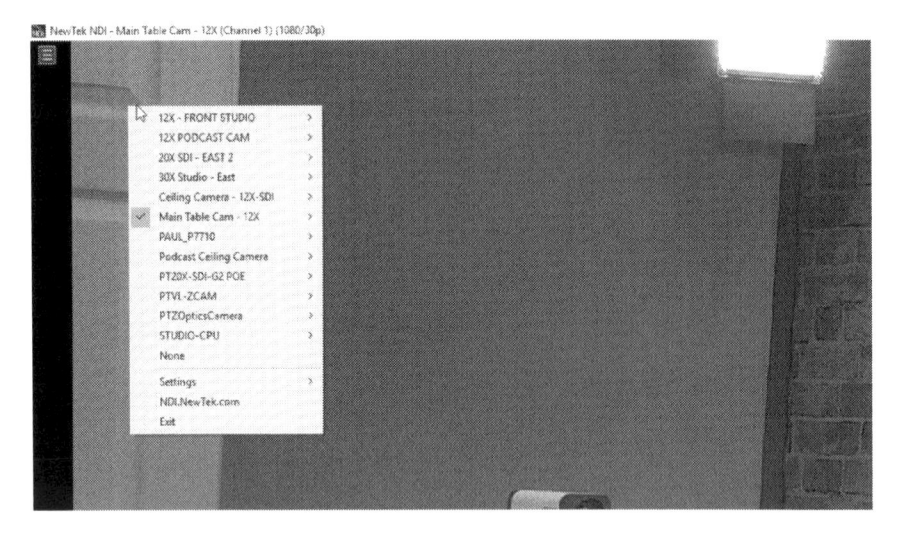

Now that you have a good understanding of how to use NDI, let's bring an NDI camera into OBS. Most NDI cameras will use an ethernet cable or WiFi to connect to OBS over your network. This eliminates the need for expensive capture cards. To add NDI functionality to OBS you need to download and install the plugin. Just search the OBS Forum for NDI and make sure you download the latest version of the plugin. You should also download the latest NDI tool set as well.

To set up an NDI camera, click on the plus button under sources, and choose "**NDI Source**." The great thing about NDI is that it will automatically discover all NDI sources available on your network. So, when you click the dropdown menu, you will see a selection of NDI sources available on your network. This step assumes that you already have an NDI camera available on your network. An easy way to test this out is to download one of the NDI camera applications available for iOS or Android. Assuming that your smartphone is connected to your WiFi, you should see your smartphone available as a camera source in the OBS NDI menu. It's as simple as that!

OBS also features the ability to output NDI video to another computer or system on your network. You can also use a new filter that allows you to make individual sources NDI outputs as well.

Now let's learn how to use your NDI video sources to power video screens and take control of remote PTZ cameras. The NDI Studio Monitor is an essential tool for video production professionals because it can be installed on any Mac or PC computer to display any NDI source. Studio Monitor is ideal for camera operators who want to gain a low-latency video preview of any NDI camera located throughout their facility. Studio Monitor is also suitable for easily displaying a video output from your NDI compatible video production software such as OBS, Wirecast, vMix, Livestream Studio, MimoLive, and many more. Let's start by looking at the NewTek NDI Studio Monitor as a remote camera operator.

As an NDI camera operator, you will almost certainly be remotely controlling your cameras over the network. You can get started by opening the NDI Studio Monitor and right clicking on the screen. This will open all of the NDI Studio Monitor features. You will notice that a list of all NDI sources will open, where you can select the NDI source you would like to display. If this NDI source is a PTZ camera, you will be presented with multiple options for remote PTZ camera control. If you have a USB Xbox joystick controller, you can connect it to your computer and enable it for use in the settings area under "**PTZ Controls**." A little-known feature about the NDI Studio monitor is the built-in hotkeys for PTZ camera operation. You can use the following hotkeys as a camera operator to improve your workflow.

NDI Studio Monitor Hotkeys

Arrow keys : Pan, Tilt the camera.*
+/- : Zoom in, out.*
Page Up, Page Down : Focus in, out.*
Home/End : Exposure up, down.*
F : Toggle auto-focus on or off
E : Toggle auto-exposure on or off
1-9 : Recall preset.
Ctrl + 1-9 : Store preset.

*Hold CTRL key for higher precision control.

If you would like to access more advanced camera settings, you can click the cog in the right-hand side of your screen at any time. You can record video using Studio Monitor by clicking the red record button in the bottom left-hand side of the application. Our testing shows that Studio Monitor is only capable of recording NDI HX (High Efficiency) video feeds directly to the hard drive. This may change in a future update. This feature allows remote camera operators to record high-quality videos directly to their local hard drive. These recordings can then be used in post-production or for video analysis. For example, a sports coach could easily set up a picture in picture using two cameras. One PTZOptics 30x camera could be set up to capture up-close details and a wide angle PTZOptics ZCam Camera to capture the entire field. These recordings can then be used for presentations with coaches and athletes. Coaches can even use an NDI telestrator with a touch screen tablet to highlight plays just like you see in professional sports.

The next powerful use for the NDI Studio Monitor is for previewing video outputs. The NDI Studio Monitor is ideal for extending your camera and video production outputs to multiple video screens. These could be located in lobbies, building façades, or on outdoor billboard display screens. Providing video sources for these areas over traditional analog cabling was always impractical and HDMI distance limitations is a headache many of us have dealt with in the past. IP is, of course, the perfect solution because ethernet cabling can be run hundreds of feet and it can be plugged into a simple network switch.

So, there you have it. NDI in OBS and NDI Studio Monitor are powerful tools for video production and the NDI ecosystem.

Section 3, Lecture 21: Using an NDI Telestrator in OBS

Another great NDI tool is called the NDI Telestrator. Just like any other NDI source, assuming you have an NDI Telestrator running on your network, it will be available to OBS. But before we start, what the heck is a Telestrator? It is a device or an application that allows you to draw freehand sketches or diagrams over a video using an alpha channel layer. You have probably seen it used on TV, by presenters and analysts when they make freehand sketches on the screen, especially during sports analysis or when explaining weather patterns.

There are a couple of NDI Telestrator options available today including one from NewTek and another from Panamation. On the Telestrator device, you have a drop-down menu at the top of the screen that automatically pulls in all of your available NDI sources. The Telestrator allows you to select the source you want and annotate on top of it. Not only can you annotate on that source, but you can also bring the Telestrator into the video as a different source as well. Additionally, the Telestrator offers various options for making annotations on the video. You can select between colors, shapes, highlights, and more. One of its best features, though, is the ability to take a picture of something happening on the video and annotate on the resulting picture to explain something to your viewers. This is a great way to enhance presentations and many other video productions with live action.

Section 3, Lecture 22: OBS Instant Replay - Video Buffer

Have you ever wanted to go back in time? Did you know OBS offers an instant replay option? This is one of the lesser known features of the software. The OBS instant replay feature can save recordings of your video production in your computer's RAM (Random Access Memory). OBS can record and replay backup to 8 gigabytes of recorded video. You can set up OBS instant replay in the "**Outpu**t" tab of the main OBS "**Settings**" area. In the advanced section of the output tab, you will find the "**Replay Buffer**" tab which is where you can enable instant replay. This is where you will have two variables to adjust with regards to your instant replay recording. You can adjust the "**Maximum Replay Time**" which will affect your "**Estimated memory usage.**" The location and quality of these recordings will be reflected by the settings you are using in the "**Recording**" tab.

Note: It's not suggested to use more than 50% of your total available RAM for instant replay. You can check out your available RAM utilizing an application like Windows Task Manager.

Setting up the replay buffer adds a new button in the "**Controls**" box on the OBS dashboard. You should now have a button that says "**Start Replay Buffer**" right above the "**Studio Mode**" button. OBS does require you to also set up a hotkey for "**Save Replay**" to use this feature. Once the replay buffer is set up in OBS, it's time to review where your recordings are being stored.

By default, your instant replay recordings will be saved in the same folder you have set up for normal recordings, except each file will have a custom prefix/suffix which you can set up in the "**Advanced**" settings area. The "**Advanced**" settings area has a "**Recording**" section where you can adjust this information and select whether you would like to "**Overwrite if file exists.**" When you choose "**Overwrite if file exists**" you are essentially setting up a single file that you can recall repeatedly for use as an instant replay. Next give your "**Filename Formatting**" area a static name such as "instant-replay" and click apply. Now that you have your naming conventions all set up, you can create a file using the replay buffer, so that you can reference this files source in a new scene.

You can now add a new scene called "Instant Replay" with a **"Media Source"** that maps directly to your instant replay file in the folder you have set up. Now you can transition to this instant replay scene whenever you want to show a recording you have made quickly. Note that it can take two to three seconds for your replay buffer file to finish recording. Therefore, this is the perfect chance to properly use a three-to-five second stinger transition used with this scene. The three-to-five seconds a stinger transition takes to complete should give your video enough time to fully render and be ready for playback.

The instant replay feature is mostly used for sports (and eSports).

Section 3, Lecture 23: PTZOptics Camera Control with an Xbox Joystick

As you can start to see, the Open Broadcast Software platform can handle advanced video production setups for those who understand how to use it. Now that you have an understanding of the NewTek NDI, you can use this knowledge to use PTZOptics cameras in OBS using a plugin to control the cameras with an Xbox Joystick. PTZOptics is a camera manufacturer who has supported the OBSProject for many years. The company has released an OBS plugin for users that is an excellent example of extending the functionality of the OBS. The plugin allows users to take control of PTZ cameras inside of OBS quickly and in the tutorial, it is shown how to use HTTP commands inside OBS to automatically call PTZ camera presets for a camera when you click on a scene.

A "PTZ" or pan, tilt, and zoom camera preset is a saved camera position stored inside a PTZ camera. This "PTZ Preset" can be used to quickly move a robotic camera to a known location in your studio or on stage. PTZOptics released a plugin that allowed live streamers to control their PTZOptics camera directly from inside OBS. PTZOptics has continued to listen to the OBS community and make improvements to the plugin, and it now features Xbox joystick support and also hotkeys support. Using only an Xbox joystick, plugged into your computer via USB, you can fully control PTZOptics robotic cameras over an ethernet connection on your local area network.

Using an Xbox joystick inside OBS to control your PTZOptics camera is quite simple and powerful for many live steamers. After you have the plugin installed, go to the main menu on the OBS dashboard and click on "**TOOLS**" to find the new "**PTZOptics controller.**" Clicking on this will bring up the PTZOptics Control pane, and you can immediately see that it offers complete control for up to eight cameras. To set up the Xbox joystick with OBS, click the settings tab at the top of the PTZOptics Control pane and select "**Joystick Settings**." A small box opens up and, if you have an Xbox joystick connected, it will be indicated inside the box. Now all you have to do is check the square that says, "Use Joystick" and click okay. That connects the joystick to OBS and your PTZOptics cameras.

Additionally, inside the PTZOptics Control pane, you have buttons to store camera presets. Some OBS users have reported that it is easier to build PTZ camera controls into each scene using a **"Browser"** input. To do this, you can use the new HTTP-CGI Command Sheet from PTZOptics. This allows you to select a command and build it into your OBS scene via the web browser input. You can download the HTTP-CGI Command Sheet here: PTZOptics.com/Downloads. This is an exciting example of extending the power of OBS with a video-production plugin.

5 CREATIVE WAYS PEOPLE ARE USING OBS AND LIVE VIDEO

Uses in Education

Live streaming is proving to be a particularly useful tool for instruction and training. It has expanded student learning beyond the limits of classroom locations. Teachers are using online video to enhance learning and encourage student collaboration beyond the traditional brick-and-mortar classroom (Wolf, n.d.). The "flipped classroom" is a popular new form of blended learning that encourages students to learn from online video lectures at home and complete the traditional "homework" assignments in the classroom. Here are a few ways that educators are employing live streaming:

Learning Management Systems & Online Tutoring

For over two decades, schools and educators have been using online tutoring to expand their ability to reach students beyond the four walls of a classroom. Many parents who prefer homeschooling are adopting online tutoring systems to enrich their children's education experience. Beyond this, traditional schools are using video to extend instructions to children beyond school hours and during vacations ("Why You Need to Start Thinking About Mobile Learning," n.d.).

Virtual field trips

Due to budget cuts, schools are continually searching for ways to save money without altering the quality of education that they offer to their students. As highlighted by educationworld.com, one area most affected by cuts in educational spending is the field trip. Additionally, many parents are reluctant to allow their children to go on field trips due to concerns for their safety. Teachers may visit locations and live stream the content back to their classrooms, where students gather around large screens to watch and be part of the experience. There are also companies that now offer a flipped version of the field trip to schools—instead of going to the location; the location comes to the students via live video ("Get Outta Class with Virtual Field Trips | Education World," n.d.).

Morning Announcements

Live video allows more people to participate in school events, and it increases communications. Parents, relatives, or even friends who are unable to attend plays, recitals, graduation ceremonies, and other events in-person can log into the live video stream and be a part of the community. Many schools offer an extracurricular broadcast club for students to be a part of. These clubs will often deliver a morning-announcements show for the school to watch each week. Check out my new book called the "Accelerated Broadcast Club Curriculum" to learn more on this topic.

School Sports

Schools are often using live streaming to broadcast sports events to an even wider audience than they would have reached in the past. Many sports teams will review live sports recordings to help improve athletic training and competitive analysis. With the increasing number of school sporting games now available online, college scouts are reviewing the footage to find athletes. Students are actively sharing their best performances with college scouts to help garner scholarship opportunities.

In Business

Webinars, Seminars, Courses, and Live Training

The use of live video for educational purposes is not limited to educational organizations alone. Companies and individuals whose businesses could benefit from educating customers about their products are employing live webinars with exceptional success. Offering training online, live, and by video provides advantages for both the instructor and the students. It cuts the cost of travel to and from locations. Every participant can log in from a convenient location. And it feels like in-person instruction because the students and instructors can interact via the live chat and comments section of the video. Companies are also using live video training to significant effect, especially organizations that have hundreds of workers and representatives distributed across a vast geography.

Onboarding and Employee Training

Companies and organizations are using live video to bring employees up to speed on private events happening inside a company. Video is being used for on-boarding of new products, services, policy updates, and changes in the organization. Video is helping to reduce the expense of training staff and having trainers visit multiple locations and offices. Live video allows the same content to be delivered to everyone in the organization, across geographical boundaries, at the same time. When used as a tool for providing training, students can learn from a familiar and comfortable environment and also communicate with the trainers easily, via chat. The convenience of being able to log into live streaming sessions from any location makes it ideal for busy executives and employees (Kishore, 2018; "Live Streaming," n.d.).

Customer Support

Businesses can use OBS to deliver a personalized response to customers' issues and queries via video. Many common customer support issues can be avoided with simple instructional videos designed for those who would otherwise contact your business with a problem or complaint. Creating niche sales, marketing, and support video content is ideal for companies that sell physical or virtual products that users may find complicated to use. Companies that are implementing this kind of support are experiencing considerable boosts in their customer reviews due to significantly improved customer satisfaction (IMPACT, 2019).

Product Launches and Promotion Kickoffs

A live video event is one of the best ways to engage with loyal customers and your organization's followers on social media. Businesses use the "buzz" from such events to achieve a snowballing effect across the web that can accelerate product marketing. Companies can also gain traction when they incorporate incentives with these launches—such as offering limited versions of a product that can only be obtained via the live stream. Amazon is now allowing merchants on the platform to promote sales with live video in this way. These kinds of events offer opportunities for consumers to directly interact with the people who make the products and services that they use. When used correctly, video can quickly mitigate common consumer purchasing objections and build brand equity.

Product/Service Demos and Tours

If your business sells a product that can be put to different uses, live presentations can be used to educate customers and prospects. It is also useful for showing updates to new versions of existing products, like software upgrades. Companies may also stream regularly by scheduling broadcasts to particular days and times, so their audience always knows when to log in. For example, if your company sells makeup, you could create a live video program where you show how to apply makeup to achieve specific looks and effects. To make the videos more exciting and relevant to the audience's needs, companies that employ this strategy have found it helpful to ask their followers to send in topics they would like to have discussed. As with most live video events, offering incentives like gifts or small prizes can increase the number of people who log in to watch. Or better still, consider rewarding those who are loyal to your brand or who have become brand advocates, by periodically featuring them on the show alongside the regular host. Organizations can also use live video to organize office tours and behind-the-scenes access for customers to see how products are made (Agrawal, 2019).

Q&A Sessions

A live video Q&A session is an excellent way for companies to educate and solve the problems of current customers and potential buyers. Customers are always eager for a chance to connect with the people behind the brands that they use. Those interactions offer a goldmine of opportunities for companies because, through them, customers can bring end user perspectives to light. Very often, users will tell a company exactly how they are using their products and services. Users will offer suggestions on product improvements, which can give businesses valuable insights into what their customers want and how they can better meet their needs. A live Q&A session carries a level of authenticity that cannot be achieved via the website's FAQ and allows organizations to align their brands more closely with their customers.

Live Video Contest

Another proven way to boost your company's marketing efforts is to create a contest via live video. During the premiere of "Mr. Robot," the USA Network used this strategy to significant effect. The network created a three-day live streaming event on the biggest social-gaming platform in the world, Twitch.tv. The event won them the Shorty Award for best live-contest promotion. There are many ways businesses can organize contests via a live stream.

News, Sports and Entertainment

Live News

"By leveraging video streaming tech, amateur journalists are shaking things up in the media—and by all accounts, this is only the tip of the iceberg," Julian Dossett wrote in a September 19, 2018 article on prnewswire.com. News companies and freelance reporters are making a difference with audiences through live video streams of events on location. Affordable but highly-advanced technology is allowing ordinary people to broadcast momentous events in the lives of individuals and nations from remote areas around the world. Live streaming technology is also enabling live-reporting on events where it did not use to be possible, such as one journalist's eight-hour broadcast of an incoming hurricane ("Journalism in an Instant," 2018).

Events

Organizers, as well as participants at events, are finding great uses for live streaming. Speakers at conferences and seminars have perfect opportunities for live streaming. Apart from events that are organized by big companies, individuals can also get in on the action by streaming their own personal events, such as weddings and parties. Some businesses have used live-streaming events to attract massive attention to their brands. A popular example would be the Tough Mudder's live stream of their fitness events in 2014. By 2016, the events had been viewed over 14.5 million times across various live-streaming platforms.

Other Uses

Radio and Podcasting

Yes, podcasting is audio-based but so is radio. Radio stations are expanding beyond the traditional confines of the medium by allowing listeners to experience the energy of the broadcast through live video streaming. In the same way, podcasters are letting listeners have a richer experience by enabling them to access the cast in video format by making the recording live. Some podcasters have experienced increases in their reach by putting their podcasts on Twitch, the number one social gaming platform in the world (Podigee, n.d.).

6 ORIGINS AND DEVELOPMENT HISTORY OF OBS

History

Open Broadcaster Software started in 2012. It is a free and open-source software for recording and streaming video. OBS is written in C and C++. The software provides users with a efficient alternative to paid software. OBS was born out of its author's own frustrations. In 2012, Hugh "Jim" Bailey wanted to stream Starcraft and other games to his friends online. Somebody told him about an application he could use to do that. But when he discovered he needed a subscription to use the app, he decided to build his own and make it free for anyone to use. What came out of that decision was Open Broadcaster Software.

In the author's own words:

"When I first had an interest in streaming Starcraft and games for my friends to see, I heard about a certain other app out there, and I thought, "A subscription? Seriously? I could probably write this myself" . . . and that's exactly what I did. So, two or three months later, after much learning and much toil, I have completed the first public alpha version of my streaming application. This application supports capture cards, webcams, as well as software desktop capture. You can make scenes, bitmap overlays, and it even has a plugin API so developers can add their own functionality to the application. It's fairly simple in design, and relatively easy to use. It's also written entirely in C/C++ and Direct3D 11 to maximize performance. Best part is it's entirely open source and free." ("I made a streaming application so I could stream Starcraft. Now it's open source and free for everyone. Starcraft," n.d.)

Since that first version, created in 2012, OBS has taken a life of its own and has grown far beyond the small project for live streaming games conceived by its author. In his initial post on Reddit (https://www.reddit.com/r/starcraft/comments/z58e9/i_made_a_streaming_application_so_i_could_stream/) inside the Starcraft section, Bailey asked the community to test the software and report back any bugs or issues they discovered. The response was immediate and overwhelming. From that point, OBS took off in a big way.

With the help of an army of online collaborators, development of improved versions of the software began in 2014. The work focused on producing a rewritten version of OBS to be called OBS Multiplatform. OBS Multiplatform was designed to offer multiplatform support, a more robust feature set, and a more efficient API. OBS Multiplatform was later renamed OBS Studio.

OBS has gone through many iterations and updates over the years. Always make sure to "check for updates" and read the included developer notes to keep up with new feature additions.

7 POPULAR TOPICS ON THE OBS FORUM AND TOP OBS PLUGINS

What are the discussions getting the most love on the OBS forum? The forum is an open meeting place for the OBS community. Within it, you can find help, support, and instructions for OBS.

The OBS Forums can be found here: https://obsproject.com/forum/.

Hot Topics Inside the OBS Forum:

The Most Popular Topics *(OBS Studio)*

- How to Set Up Your Own Private RTMP Server
- Setting Up Live YouTube Chat That Updates with New Streams
- How to Remove/Delete Sources Manually
- Seamless Way to Setup a Second Streaming PC
- Color Space, Color Format, and Color Range settings Guide
- Android Webcam Guide for OBS
- Installing Plugins for OBS and OBS-Studio

NewTek NDI™ Integration into OBS Studio 4.5.3

This plugin provides video input and output for OBS Studio over a network, without the need for an expensive capture card. It works with NewTek's NDI technology to add a simple audio/video input and output over IP. You can capture video at very high quality with low latency. The plugin has the following integrations:

- NDI Source: add NDI Sources into OBS like any traditional source.
- NDI Output: transmit the main program view over NDI.
- NDI Filter: a special OBS filter that outputs its parent OBS source to NDI.

With this plugin installed, OBS Studio can share video and audio between any NDI-enabled product on a local area network in real time ("obs-ndi - NewTek NDI™ integration into OBS Studio," 2019).

Advanced Scene Switcher 1.4.0

This plugin will let you automatically switch to specified scenes during streams.

Stream Effects 0.5.0pr3

With this plugin, you can liven up your screen with modern effects! Stream Effects lets you blur out parts of the screen in real-time (to hide unwanted content) and 3D Transform (A filter that you can use to move, rotate, scale and shear source in 3D space) ("Stream Effects | Open Broadcaster Software," 2019).

Stream/Recording Start/Stop Beep (SRBeep) 1.1.1

This is a handy plugin in that it does a small but essential thing. It lets you know when you actually start recording and when you stop. It is very useful for times when you think you are live but there is really nothing going on. It will notify you if your recording stops before you want it to. It works by playing a small sound to alert you when recordings start and stop. It is simple, quick to install, and super-useful ("Stream/Recording Start/Stop Beep (SRBeep)," 2019).

PTZOptics Camera Controller for OBS 2018-04-30

This plugin allows you to control your PTZOptics camera right inside OBS ("PTZOptics Camera Controller for OBS," n.d.).

Other Popular Plugin

- Remote control of OBS Studio made easy 4.5.0;
- OBS-VirtualCam 2.0.2;
- Linux Browser;
- iOS Camera for OBS Studio 2.4.0;
- Browser Plugin;
- Logitech LCD Plugin 1.

8 AUDIENCE ENGAGEMENT AND COMMUNITY BUILDING OPPORTUNITIES FOR LIVE STREAMING

Live Streaming

There are three things that you need to make your live stream successful:

- Quality content

- An audience

- Interaction and engagement

In this chapter, we will look at ways that you can make your stream more successful by making your videos more engaging.

Audience-Building and Engagement

Relevance

How up to date the things you talk about will influence how much interest your viewers will invest in your broadcasts. For a business, it will be very difficult to get people to sign in to watch a live stream, if all you ever do is try to sell them something. You should try to align your own goals with those of your live viewers. If your brand positions itself as sharing goals of your customers, your products and services will be perceived as more valuable by consumers ("Storytelling to make an impact - YouTube," n.d.).

Brevity and Confidence

There is no predetermined duration for a broadcast that will hold people's attention. Although short and to the point is the standard rule, a long video with content that speaks to people will work better than a short one that is sparse on ideas. Facebook's guideline says: "We recommend that you go live for at least 10 minutes," while livestreamingpros.com recommends a minimum of twenty minutes. It takes a bit of intuition to gain an understanding of when to "wrap things up." If you know your content, you should have a good idea of how much time you need to deliver it. You do not want to lose your audience midway because you got long-winded and you also do not wish to terminate the broadcast when you are just beginning to pique viewers' interest. Do not use fill-up time with useless content and never wander from your subject too far. But most importantly, remember that you are live; your body language is probably the most significant factor that will determine how well people receive you ("Facebook Live Tips | Create Engaging Live Videos," n.d.; "How long should your live streams be?" 2016).

Identify A Niche

Your videos should have an identity. If you are an educational institution, people should know exactly what age levels you are focused on. If you are a public speaker, your viewers ought to know what problems you are trying to solve for them. By creating a consistent focus with your messages, you will be able to build a community around that message. Knowing your product and knowing who your product is for will help you save valuable time, money, and effort chasing the wrong crowd. Even if you have a service or product that appeals to people across different demographics, it is usually better to start by focusing on one or two niches and then branching out into others as your reach increases.

Have A Consistent Streaming Schedule

People soon forget you if you do not stream regularly. Depending on the nature of what you are streaming, one day in a week or once a month may be sufficient. When dealing with younger audiences, more often is better. Regardless of the specifics of your target viewers, consistency is essential. Do not stream a fantastic video this week and then disappear when you are just starting to build momentum. Consistency is especially important at the beginning stages of audience building ("What consistency can do for your brand, with Vincenzo Landino," 2016).

Let Your Viewers Contribute Topics

Ask your viewers questions. Question what viewers may have always assumed to be true. Note that your live stream does not always have to be directly connected to your business. A business selling furniture may decide to do regular live streams where it talks about the struggles of home decor. Although not directly related to its business, that business can build a buzz around its content because it resonates with viewers. People will like their page, and as more people come in to watch their videos, some of that audience will also take an interest in the furniture that the company sells. By letting viewers send in things they would like to see, you will be increasing the appeal and impact of your streams because your audience is not only engaged but participating. Check out this diagram explaining the way LinkedIn live streams can go viral.

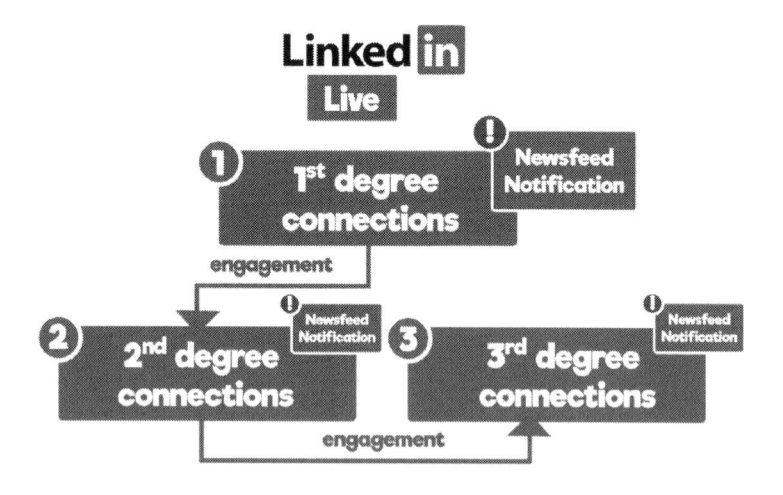

Have A Day for Live Q&A

You never know what questions people have at the back of their minds about you and what you do. Holding a live Q&A session is one way to find out. The spontaneity of the questions and the fact that you have to answer "off the cuff" makes such interactions very authentic. People often leave live Q&A sessions feeling that they know you better. You can think about live viewers as people who are beginning to view you as their friend.

Top Platforms to Build Your audience

The platform you choose for streaming your video is essential. Below is a list of the top platforms:

Facebook

You are probably already on this platform, and so are the people you want to reach. So, it makes sense to consider this as the first place to start streaming. However, you need to know that Facebook controls the reach of live broadcasts through its newsfeed-ranking algorithm. Only a percentage of your followers will be notified of your live video. Facebook is becoming less valuable as an organic tool for reaching people and more important as a paid advertising vehicle. One major drawback to Facebook is that it is a members only platform and some people may not be able to view your Facebook live stream without an account.

Pro Video Tip

Facebook is a great place to learn about your audience. Facebook will tell you which demographics are responding best to your content.

YouTube Live

YouTube is great for furthering relationships with customers, engaging with prospects, and answering questions via a live chat. Additionally, your videos can be posted to your YouTube channel after your live streams with zero delay. YouTube is the world's second largest search engine, and it can be used very effectively as a marketing tool when used in combination with Google Adwords. YouTube is often preferred to Facebook for public events because no log-in is required.

Instagram

This platform is ideal for fashion, makeup, and the entertainment industry. Instagram is primarily a photo-sharing platform, and therefore, audiences enjoy interacting with businesses that use a lot of photography.

Vimeo

Businesses can use Vimeo to live-stream events. One of the essential things about Vimeo is that streamers can use their professional tools for monetization. Vimeo can be used to charge viewers for access to premium events that are either live or available on-demand.

Twitch

Twitch is quickly becoming a favorite destination for live streamers because the platform has a sort of magnetic feel. While Twitch is most popular with young men, women demographics are growing on the platform. The platform has perhaps the most attractive monetization system in place to pay out broadcasters for their content. The platform features broadcaster extensions which allow new levels of interactivity with viewers not seen on any other platform.

Strategies for Interacting with Your Audience

At StreamGeeks, our mission is to help businesses discover the power of live streaming. We see live streaming as the bridge between internal teams and external communities on social media. Using this type of video communication can be an incredible differentiator for your business. We have found that real-time engagements with your audience help create an authentic form of communication, unlike any other type of media. Facebook recently released a report showing that viewers spend three times as much time watching live videos as they do with on-demand video content. Forbes Magazine recently published a statement noting that customers are 64% more likely to purchase a product after watching a video (Forbes, 2017).
At StreamGeeks, we have a simple recipe for success:
Great Content + Live Viewer Engagement + Social Media Advertising = Success.

As a good marketer, your team surely has the goal of attracting new subscribers, page likes, and viewers for your content. While this is always a great place to start, it's essential to think about the customer journey when focusing your live streaming and video marketing.

Attract Stage

In the early attract stage of your funnel, it's essential to grow your social media audience with Facebook and YouTube engagements. Because live streams are so exciting to viewers, you'll find that it performs well attracting strangers to subscribe to your social media channels.

Always give your viewers a compelling reason to subscribe to your channel right before you ask for the all important like or subscribe click. For example, at the end of our videos, we will say: "There are a lot of great ways to leverage live streaming for your business. We create new suggestions every week, so subscribe to our channel to stay up-to-date with our latest content."

Convert Stage

Ultimately, your goal is to create leads. In the convert stage, try to engage your audience in the live chat room. You can ask your audience questions during the live broadcast that prompt responses and increase engagement. This is your chance to casually learn a little bit about your audience and identify your most-engaged viewers.

Further along in the conversion stage, you can offer a "content upgrade"—such as an e-Book or guide. This is the crucial next step for your audience as they enter into your CRM. During this step, use landing pages and forms to automate the process. Once a contact fills out a form, you can have an email sent automatically to the correct person or team to follow up.

Close Stage

In the close stage of your video marketing funnel, you want to deliver live content that is appropriate for a contact who has already expressed interest in your offerings. Popular live video content in the close stage include promotional events, QVC-style videos, and live testimonials.

Live testimonials are a great way to present customer success stories and to allow current prospects to ask questions throughout the broadcast. Clients willing to come on screen for a live testimonial will make an impactful impression on your audience.

Delight Stage

Finally, the delight stage is perhaps the most crucial step for retaining your customers in the long run. Here you can use live streaming for community building and leverage social media as a tool where your live viewers can discuss topics in more detail with like-minded professionals.

Facebook has recently built in a new feature for live streaming privately directly to Facebook groups. This is a great way to offer exclusive community content in the delight stage for live video marketing.

The Live Video Marketing Funnel

In each stage of your lead funnel, it's important to gear the material towards the buyers' next step in their journey. When it comes to video marketing, you can apply the same thought process. Video marketing and live video marketing are not that different when it comes to strategy.

In the attract stage of the video marketing funnel, you may be creating brand videos and uploading them to YouTube. In the same stage of the live video marketing funnel, you may create a lifestyle stream sharing live video moments around the office.

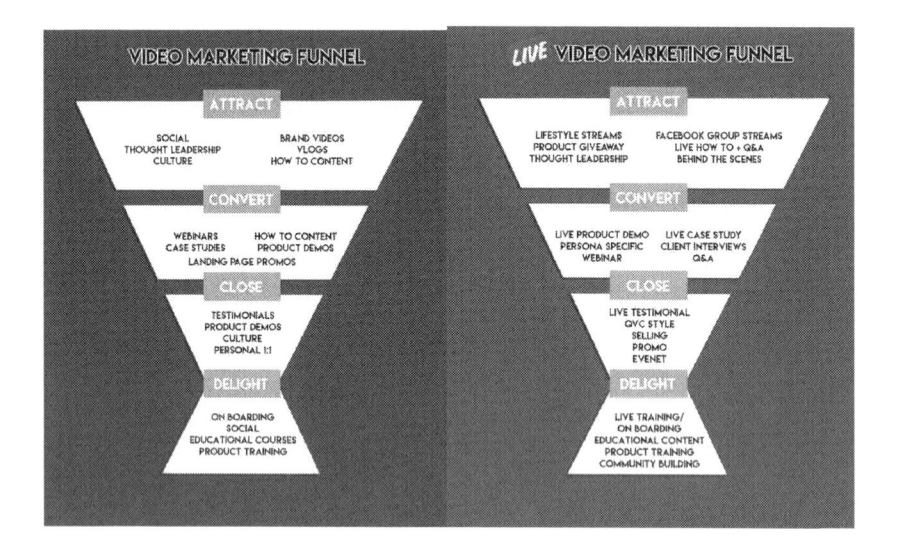

How to Implement a Live Video Marketing Process

Once you have determined an ideal topic for your live stream, it's always a great idea to think of a compelling call-to-action. So, before you schedule a live stream, be prepared with a detailed agenda and a plan ready for implementation. At StreamGeeks, our live streaming team includes two hosts and a producer. We all share a Google Doc to collaborate on before the show. With this, we can make sure everyone is on the same page about the show agenda and workflow.

Here is step-by-step process for preparing an inbound marketing campaign with live streaming:

Our Live Video Process
How the StreamGeeks go live...

- **Research**: Find a relevant topic you can provide value to your audience.
- **Determine CTA**: What is your content call to action?
- **Schedule Live Event**: Schedule our event on both YouTube & Facebook.
- **Giveaway Contest**: Use HubSpot Lead Flows to drive extra traffic.
- **Schedule Email**: Schedule email for 10 minute before live show to target personas.
- **Test Equipment**: Always start testing AV equipment before live show.
- **Start the Pre-Show**: It's great to have a 5-10 minute preshow to warm up the audience.
- **Fire & Launch**: Start your live show!

1. Research your topic.

Brainstorming is a fun part of marketing, and you can use social media to help your team determine what could work for your next video. Once you have a social media following that cares about the content you're creating, you can engage your online community and effectively crowdsource the ideas that are most important to your users.

If you have a bunch of ideas and aren't sure which one will have the best impact, consider searching through Google's trending tool. You can use this tool to see where your top keywords connect with popular search phrases that may be trending up or down.

2. Determine your CTA.

Critical to your inbound lead generation success is a compelling call-to-action (CTA). A CTA is usually displayed as a lower third graphics on the bottom ⅓ of your screen during a live broadcast. This could be a live giveaway or a downloadable guide. To make your life easier during the live show, you can create a dedicated button to display your CTA.

You want your CTA to be displayed naturally during the conversation about the topic. Most live streaming software has a feature called "hotkeys" allowing your team to assign any key on a keyboard or USB device to trigger the event. Allowing your producer to quickly press the "CTA" buttons makes it easy for your team to stay on the same page and look professional.

3. Schedule your live event & email.

Once your plan is in order, you can schedule your live events on both YouTube and Facebook. You can take the direct links to these videos and insert them as CTAs in your scheduled live-show emails.

You can use your email system to optimize the delivery of your emails to the correct contact personas. Use contact personas to determine the contacts that would be most interested in the content you are creating.

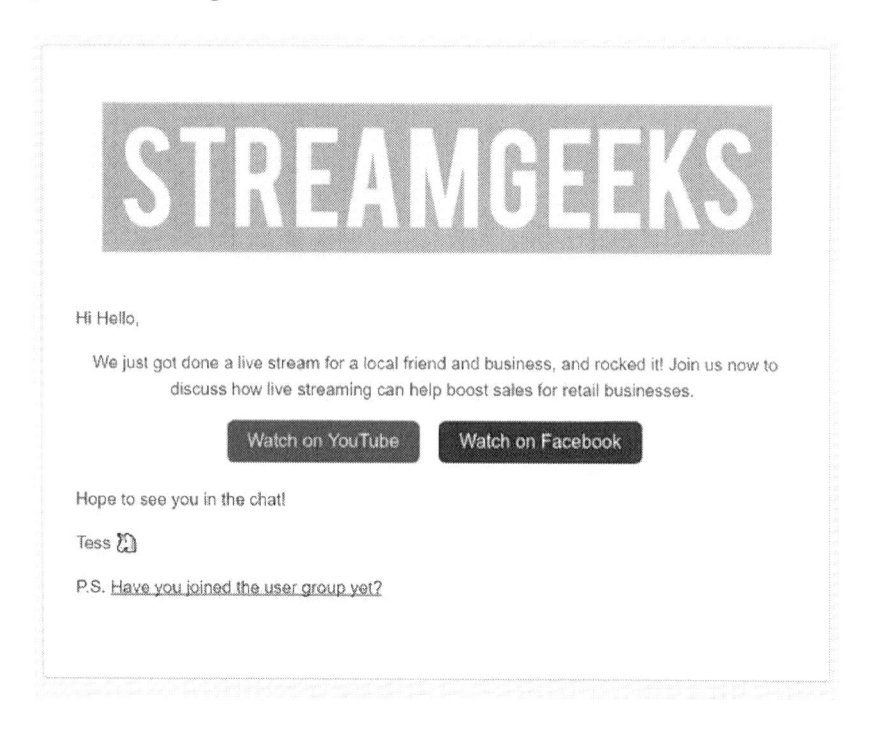

4. Schedule a live giveaway. (Optional)

If you have the resources, hosting a live giveaway is something that all audiences love. You can use Gleam.io to host your competitions and trigger social actions. These competitions can be set up to promote your social media pages and drive more live viewership to your broadcasts.

After all the planning is set, test your equipment, start your pre-show, and start the live show with a bang!

Promoting Your Live Broadcast Post-Show

Once the live broadcasts are over, you can turn the content into a blog post. Your detailed blog posts can include images from the live broadcast, the embedded video, and notes about the show. An easy way to start blogging about your video is to take snapshots from your video that you feel explain crucial steps. You can take the pictures you have collected and explain what is happening in each as a source of inspiration for your writing.

For a standard ten to twenty minute video, you can take roughly four to six pictures and post them throughout your blog post. When you are done with the blog post, you can go back to your YouTube video and paste in a couple of relevant paragraphs into the YouTube description. Let's not forget that YouTube is the world's second largest search engine.

Why Live Video is Beneficial for You

One of the wonderful benefits of live streaming is the organic reach and long-term video ranking capabilities. When a normal video is uploaded to YouTube it has to accumulate views and watch time starting at zero. Watch time is one of the most important factors that YouTube uses to rank videos. Once you have an audience willing to tune in to your broadcasts, you will be accumulating views and watch time during your live stream. As soon as your video is available on-demand on either YouTube or Facebook, it will have a huge boost over regular uploaded content. All the watch time you have accumulated during the live broadcast will automatically be attributed to your video.

11 CONCLUSION

There is something special about the live streaming and broadcast community. Unlike other careers, that are highly competitive, the broadcast industry is full of people who want to help each other. There are so many different people and perspectives; you will find that anyone can fit in with this community! Once you have read through this book and taken the online course, I highly recommend joining our OBS Users Facebook Group at https://facebook.com/groups/OBSUsers.

This book is 100% inspired by the OBS user community. I hope that this book can assist new OBS users get up to speed so that perhaps they can contribute something to the project. If anything inside this book has inspired you, share something about it on social media. I hope that the content we make here at StreamGeeks can inspire generations of broadcast professionals to join the movement!

I will leave you with a slogan we have taped to a back of LCD monitor in our studio "Get Ready and Go Live"!

Please feel free to contact me at any time via email at **paul.richards@streamgeeks.us**.

GLOSSARY OF TERMS

3.5mm Audio Cable - Male to male stereo cable, common in standard audio uses.

4K - A high definition resolution option (3840 x 2160 pixels or 4096 x 2160 pixels)

16:9 [16x9] - Aspect ratio of 9 units of height and 16 units of width. Used to describe standard HDTV, Full HD, non-HD digital television and analog widescreen television.

API [Application Program Interface]- A streaming API is a set of data a social media network uses to transmit on the web in real time. Going live directly from YouTube or Facebook uses their API.

Bandwidth - Bandwidth is measured in bits and the word "bandwidth" is used to describe the maximum data transfer rate.

Bitrate – Bitrates are used to select the data transfer size of your live stream. This is the number of bits per second that can be transmitted along a digital network.

Broadcasting - The distribution of audio or video content to a dispersed audience via any electronic mass communications medium.

Broadcast Frame Rates - Used to describe how many frames per second are captured in broadcasting. Common frame rates in broadcast include **29.97fps and 59.97 fps**.

Capture Card - A device with inputs and outputs that allow a camera to connect to a computer.

Chroma Key - A video effect that allows you to layer images and manipulate color hues [i.e. green screen] to make a subject transparent.

Cloud Based-Streaming - Streaming and video production interaction that occurs within the cloud, therefore accessible beyond a single user's computer device.

Color Matching - The process of managing color and lighting settings on multiple cameras to match their appearance.

Community Strategy - The strategy of building one's brand and product recognition by building meaningful relationships with an audience, partner, and clientele base.

Content Delivery Network [CDN] - A network of servers that deliver web-based content to an end user.

CPU [Central Processing Unit] – This is the main processor inside of your computer, and it is used to run the operating system and your live streaming software.

DAW - Digital Audio Workstation software is used to produce music. It can also be used to interface with multiple devices and other software using MIDI.

DB9 Cable - A common cable connection for camera joystick serial control.

DHCP [Dynamic Host Configuration Protocol] Router - A router with a network management protocol that dynamically sets IP addresses, so the server can communicate with its sources.

Encoder - A device or software that converts your video sources into an RTMP stream. The RTMP stream can be delivered to CDNs such as Facebook or YouTube.

FOH – Front of House is the part of your church that is open to the public. There is generally a FOH audio mix made to fill this space with the appropriate audio.

GPU – Graphics Processing Unit. This is your graphics card which is used for handling video inside your computer.

H.264 & H.265 - Common formats of video recording, compression, and delivery.

HDMI [High Definition Multimedia Interface] - A cable commonly used for transmitting audio/video.

HEVC [High Efficiency Video Coding] - H.265, is an advanced version of h.264 which promises higher efficiency but lacks the general support of h.264 among most software and hardware solutions available today.

IP [Internet Protocol] Camera/Video - A camera or video source that can send and receive information via a network & internet.

IP Control - The ability to control/connect a camera or device via a network or internet.

ISP – Internet Service Provider. This is the company that you pay monthly for your internet service. They will provide you with your internet connection and router.

Latency - The time it takes between sending a signal and the recipient receiving it.

Live Streaming - The process of sending and receiving audio and or video over the internet.

LAN [Local Area Network] - A network of computers linked together in one location.

MIDI [Musical Instrument Digital Interface] - A way to connect a sound or action to a device. (i.e. a keyboard or controller to trigger an action or sound on a stream

Multicast - Multicast is a method of sending data to multiple computers on your LAN without incurring additional bandwidth for each receiver. Multicast is very different from Unicast which is a data transport method that opens a unique stream of data between each sender and receiver. Multicast allows you to broadcast video from a single camera or live streaming computer to multiple destinations inside your church without adding the bandwidth burden on your network.

Multicorder – Also known as an "IsoCorder" is a feature of streaming software that allows the user to record raw footage from camera feed directly to your hard drive. This feature allows you to record multiple video sources at the same time.

NDI® [Network Device Interface] - Software standard developed by NewTek to enable video-compatible products to communicate, deliver, and receive broadcast quality video in high quality, low latency manner that is frame-accurate and suitable for switching in a live production environment.

NDI® Camera - A camera that allows you to send and receive video over your LAN using NDI technology.

NDI®|HX - NDI High Efficiency, optimizes NDI for limited bandwidth environments.

Network - A digital telecommunications network which allows nodes to share resources. In computer networks, computing devices exchange data with each other using connections between nodes.

Network Switch – A network switch is a networking device that connects multiple devices on a computer network using packet switching to receive, process and forward data to the destination device.

NTSC - Video standard used in North America.

OBS – Open Broadcaster Software is one of the industries most popular live streaming software solutions because it is completely free. OBS is available for Mac, PC, and Linux computers.

PAL - Analog video format commonly used outside of North America.

PCIe - Allows for high bandwidth communication between a device and the computer's motherboard. A PCIe card can installed inside a custom-built computer to provide multiple video inputs (such as HDMI or SDI).

PoE - Power over Ethernet.

PTZ - Pan, tilt, zoom.

RS-232 - Serial camera control transmission.

Router – Your internet router is generally provided to you by your internet service provider. This device may include a firewall, WiFi and/or network switch functionality. This device connects your network to the internet.

RTMP [Real Time Messaging Protocol] – Used for live streaming your video over the public internet.

RTSP [Real Time Streaming Protocol] - Network control protocol for streaming from one point to point. Generally, used for transporting video inside your local area network.

vMix® – vMix is a live streaming software built for Windows computers. It is a professional favorite with high-end features such as low latency capture, NDI support, instant replay, multi-view and much more.

Wirecast® – Wirecast is a live streaming software available for both Mac and PCs with advanced features such as five layers of overlays, lower thirds, virtual sets and much more.

xSplit® – xSplit is a live streaming software with a free and/or low monthly fee paid option. This is a great software available on for Windows computers that combines advanced features and simple to use interface.

About the Author

Paul is the Chief Streaming Officer for StreamGeeks. StreamGeeks is a group of video production experts dedicated to helping organizations discover the power of live streaming.

Every Monday, Paul and his team produce a live show in their downtown West Chester, Pennsylvania studio location. Having produced live shows as amateurs themselves, the StreamGeeks steadily worked their way to a professional level by learning from experience as they went.

Today, they have an impressive following and a tight-knit online community which they serve through consultations and live shows that continue to inspire, motivate, and inform organizations who refuse to settle for mediocrity. The show explores the ever-evolving broadcast and live-streaming market while engaging a dedicated live audience. Richards now specializes in the live streaming media industry leveraging the technology for lead generation. In his book, *Live Streaming is Smart Marketing*, Richards reveals his views on lead generation and social media.

Additional Online Courses:

Join over 20,000 other students learning how to leverage the power of live streaming! Take the following courses taught by Paul Richards.

- **Facebook Live Streaming** - *Beginner*

This course will take your through the Facebook Live basics. It has already been updated twice! This also includes using Facebook Live Reactions!

- **YouTube Live Streaming** - *Beginner*

This course will take you through the YouTube Live basics. It also includes essential branding and tips for marketing.

- **Introduction to OBS (Open Broadcaster Software)**

This course will take your through one of the world's most popular FREE live streaming software solutions. OBS is a great place to start live streaming for free!

- **Introduction to xSplit Software** - *Beginner*

This course takes you through xSplit which has more features that OBS but costs roughly $5/month. Learn how to create amazing live productions and make videos much faster with xSplit!

- **Introduction to vMix** - *Intermediate*

vMix will have you live streaming like the Pros in no time. This Windows based software will amaze even the most advanced video producers!

- **Introduction to Wirecast** - *Intermediate*

Wirecast is the preferred software for so many professional live streamers. Available for Mac or PC this is the ideal software for anyone looking for professional streaming.

- **Introduction to NewTek NDI** - *Intermediate*

NewTek's innovative IP video standard NDI (Network Device Interface) will change the way you think about live video production. Learn how to use this innovative new technology for live streaming and video production system design.

- **Introduction to live streaming course** - *Beginner*

This course includes everything you need to get started designing your show. This course includes a starter pack of course files including: Photoshop, After Effects and free Virtual Sets.

- **Introduction to live streaming** - *Intermediate*

This course focuses on more advanced techniques for optimizing your production workflow and using compression to get the most out of your processor. This course includes files for: Photoshop, After Effects and free Virtual Sets.

- **Live Streaming for Good - Church Streaming Course** - *Intermediate*

This course focuses on live streaming for churches and houses of worship. We tackle some of the big questions about live streaming in a house of worship and dive into the specific challenges of this space.

- **How to Live Streaming A Wedding** - *Beginner*

This is a great course for anyone looking to start live streaming weddings. Originally designed for Wedding Photographers to add a live streaming service to their existing portfolio of offerings. This course is great for beginnesr

References

1. "Professional Development for Primary, Secondary & University Educators/Administrators." (n.d.). Retrieved March 8, 2019, from http://visualteachingalliance.com/
2. "5 best and worst things about the Mevo Livestream Camera – FilmIn5D." (n.d.). Retrieved March 10, 2019, from http://filmin5d.com/5-best-and-worst-things-about-the-mevo-livestream-camera/
3. (5) "How to Create a Tribe Around Your Brand" | LinkedIn. (n.d.). Retrieved March 10, 2019, from https://www.linkedin.com/pulse/how-create-tribe-around-your-brand-anup-batra/
4. (48)" WLR FM Is Number One In Waterford City and County!" - YouTube. (n.d.). Retrieved March 8, 2019, from https://www.youtube.com/watch?v=4TKRtuJEC18
5. "Advanced Scene Switcher." (n.d.). Retrieved March 10, 2019, from https://obsproject.com/forum/resources/advanced-scene-switcher.395/
6. Agrawal, A. J. (n.d.). "Why Live Streaming Should Be a Part of Your Marketing Plan." Retrieved March 10, 2019, from https://www.forbes.com/sites/ajagrawal/2016/06/01/why-live-streaming-should-be-a-part-of-your-marketing-plan/
7. "Alexa Top 500 Global Sites." (n.d.). Retrieved March 8, 2019, from https://www.alexa.com/topsites
8. "Android Webcam guide for OBS." (n.d.). Retrieved March 10, 2019, from https://obsproject.com/forum/resources/android-webcam-guide-for-obs.56/
9. Ave, I. E., & Suite 102Clevel. (2018, January 10). "28 Video Stats for 2018." Retrieved March 8, 2019, from https://www.insivia.com/28-video-stats-2018/
10. "Best streaming software in 2018: A breakdown of the main contenders." (2018, January 10). Retrieved March 10, 2019, from https://www.epiphan.com/blog/best-streaming-software-2018/
11. Bowman, M. (n.d.). "Video Marketing: The Future of Content Marketing." Retrieved March 8, 2019, from https://www.forbes.com/sites/forbesagencycouncil/2017/02/03/video-marketing-the-future-of-content-marketing/
12. BruBearBaby. (n.d.). *Baby Laughing Hysterically at Ripping Paper (Original)*. Retrieved from https://www.youtube.com/watch?v=RP4abiHdQpc
13. Christopher Wade. (n.d.). *Wisdom Teeth Wake Up at the Dentist*. Retrieved from https://www.youtube.com/watch?v=vyR2KNNsxCc&gl=NG&hl=en-GB
14. "Cisco Visual Networking Index: Forecast and Trends, 2017–2022 White Paper." (n.d.-a). Retrieved March 8, 2019, from https://www.cisco.com/c/en/us/solutions/collateral/service-provider/visual-networking-index-vni/white-paper-c11-741490.html
15. "Cisco Visual Networking Index: Forecast and Trends, 2017–2022 White Paper." (n.d.-b). Retrieved March 8, 2019, from https://www.cisco.com/c/en/us/solutions/collateral/service-provider/visual-networking-index-vni/white-paper-c11-741490.html
16. Clarine, A. B. (2016, September 29). "11 Reasons Why Video Is Better Than Any Other Medium." Retrieved March 8, 2019, from https://www.advancedwebranking.com/blog/11-reasons-why-video-is-better/
17. CNN.com, S. C. S. C. has been writing about tech since 2000, appearing in publications such as: World, P. C., InfoWord, & Others, M. (n.d.). "What Is Streaming and When Do You Use It?" Retrieved March 8, 2019, from https://www.lifewire.com/internet-streaming-how-it-works-1999513
18. "Earn monetization revenue though Facebook Livestreaming." (2 years ago). Retrieved March 10, 2019, from https://vidooly.com/blog/earn-monetization-revenue-though-facebook-livestreaming/
19. "Easily integrate Skype calls into your content with the new content creators feature." (2018, September 13). Retrieved March 10, 2019, from https://blogs.skype.com/news/2018/09/13/easily-integrate-skype-calls-into-your-content-with-the-new-content-creators-feature/
20. "Facebook Live Tips | Create Engaging Live Videos." (n.d.). Retrieved March 10, 2019, from http://live.fb.com/tips/
21. "Get Outta Class with Virtual Field Trips" | Education World. (n.d.). Retrieved March 10, 2019, from https://www.educationworld.com/a_tech/tech/tech071.shtml

22. Guerin, E. (n.d.). "The Top 16 Video Marketing Statistics for 2016" | Adelie Studios. Retrieved March 8, 2019, from http://www.adeliestudios.com/top-16-video-marketing-statistics-2016/
23. Hinton, S. (2017, July 7). "Lessons from my first year of live coding on Twitch." Retrieved March 10, 2019, from https://medium.freecodecamp.org/lessons-from-my-first-year-of-live-coding-on-twitch-41a32e2f41c1
24. "How long should your live streams be?" (2016, November 8). Retrieved March 10, 2019, from https://livestreamingpros.com/how-long-should-your-live-streams-be/
25. "How to Choose a Live Streaming Platform." (2018, May 23). Retrieved March 10, 2019, from https://www.streamingmedia.com/Articles/ReadArticle.aspx?ArticleID=125228
26. "How to convert FLVs to MP4 fast without re-encoding." (n.d.). Retrieved March 10, 2019, from https://obsproject.com/forum/resources/how-to-convert-flvs-to-mp4-fast-without-re-encoding.78/
27. "How to get Mevo footage into OBS – Manchester Video Limited." (n.d.). Retrieved March 10, 2019, from https://www.manchestervideo.com/2018/09/05/how-to-get-mevo-footage-into-obs/
28. "How to monetize live streaming." (2017, August 16). Retrieved March 10, 2019, from https://www.smartinsights.com/digital-marketing-platforms/video-marketing/monetize-live-streaming/
29. "How to Remove/Delete Sources Manually." (n.d.). Retrieved March 10, 2019, from https://obsproject.com/forum/resources/how-to-remove-delete-sources-manualy.718/
30. "How to set up your own private RTMP server using nginx." (n.d.). Retrieved March 10, 2019, from https://obsproject.com/forum/resources/how-to-set-up-your-own-private-rtmp-server-using-nginx.50/
31. "How to Stream to Facebook Live." (n.d.). Retrieved March 10, 2019, from https://obsproject.com/forum/resources/how-to-stream-to-facebook-live.391/
32. "How to use OBS Studio as an external encoder for Microsoft Teams and Stream Live Events." (2018, August 24). Retrieved March 10, 2019, from https://lucavitali.wordpress.com/2018/08/24/how-to-use-obs-studio-external-encoder-for-live-events/
33. "How Video Will Take Over the World." (n.d.). Retrieved March 8, 2019, from https://www.forrester.com/report/How+Video+Will+Take+Over+The+World/-/E-RES44199?isTurnHighlighting=false&highlightTerm=1.8%2520million%2520words#
34. HubSpot. (n.d.). "How to Create High-Quality Videos for Social Media." Retrieved March 8, 2019, from https://offers.hubspot.com/video-social-media-marketing
35. Hussain, A. (n.d.). "22 Eye-Opening Statistics About Sales Email Subject Lines That Affect Open Rates." [Updated for 2019]. Retrieved March 8, 2019, from https://blog.hubspot.com/sales/subject-line-stats-open-rates-slideshare
36. "I made a streaming application so I could stream startcraft. Now it's open source and free for everyone : starcraft." (n.d.). Retrieved March 10, 2019, from https://www.reddit.com/r/starcraft/comments/z58e9/i_made_a_streaming_application_so_i_could_stream/
37. IMPACT. (n.d.). "14 Ways Your Business Can Successfully Use Live-Streaming Video." Retrieved March 10, 2019, from https://www.impactbnd.com/blog/ways-to-use-live-streaming-video
38. joe-cortez. (2018, June 12). "The Importance of Being Interactive While Streaming." Retrieved March 10, 2019, from https://switchboard.live/blog/interactive-live-streaming-engagement/
39. "Journalism in an Instant: How Livestreaming is Changing Our News." (2018, September 19). Retrieved March 10, 2019, from https://mediablog.prnewswire.com/2018/09/19/journalism-in-an-instant-how-livestreaming-is-changing-our-news/
40. Kishore, K. (2018, August 10). "How Live Video Streaming Can Benefit Different Business Verticals?" Retrieved March 10, 2019, from https://hackernoon.com/how-live-video-streaming-can-benefit-different-business-verticals-9dd8bc2ed94b
41. Lagarde, E. (2018, September 24). "How to Make Money Broadcasting Live Video." Retrieved March 10, 2019, from https://www.dacast.com/blog/how-to-make-money-broadcasting-live-video/
42. "Live Streaming." (n.d.). Retrieved March 10, 2019, from https://www.panopto.com/resource/15-ways-businesses-are-using-live-streaming/
43. "Live Video Streaming: A Global Perspective." (n.d.). Retrieved March 8, 2019, from https://www.iab.com/insights/live-video-streaming-2018/

44. "Mark Zuckerberg: Within Five Years, Facebook Will Be Mostly Video | Popular Science." (n.d.). Retrieved from https://www.popsci.com/mark-zuckerberg-within-five-years-facebook-will-be-mostly-video
45. "NewTek." (2019). In *Wikipedia*. Retrieved from https://en.wikipedia.org/w/index.php?title=NewTek&oldid=877643055
46. OBS and OBS-Studio: "Install Plugins (windows)." (n.d.). Retrieved March 10, 2019, from https://obsproject.com/forum/resources/obs-and-obs-studio-install-plugins-windows.421/
47. "OBS Classic" (No Longer Supported). (n.d.). Retrieved March 10, 2019, from https://obsproject.com/forum/categories/obs-classic-no-longer-supported.37/
48. "OBS Studio 22 brings new features and improvements for streaming and recording." (n.d.). Retrieved March 10, 2019, from https://www.shacknews.com/article/106851/obs-studio-22-brings-new-features-and-improvements-for-streaming-and-recording
49. "OBS Studio: Color Space, Color Format, Color Range settings Guide. Test charts." (n.d.). Retrieved March 10, 2019, from https://obsproject.com/forum/resources/obs-studio-color-space-color-format-color-range-settings-guide-test-charts.442/
50. *OBS Studio: Free and open source software for live streaming and screen recording - obsproject/obs-studio*. (2019). C, OBS Project. Retrieved from https://github.com/obsproject/obs-studio (Original work published 2013)
51. "OBS vs Wirecast: Best Live Streaming Software?" (2017, May 24). Retrieved March 10, 2019, from https://www.videoguys.com/blog/obs-vs-wirecast-best-live-streaming-software/
52. "obs-ndi - NewTek NDI™ integration into OBS Studio." (n.d.). Retrieved March 10, 2019, from https://obsproject.com/forum/resources/obs-ndi-newtek-ndi%E2%84%A2-integration-into-obs-studio.528/
53. "OBS-Studio: High quality recording and multiple Audio Tracks." (n.d.-a). Retrieved March 10, 2019, from https://obsproject.com/forum/resources/obs-studio-high-quality-recording-and-multiple-audio-tracks.221/
54. "OBS-Studio: How to configure your Microphone—Noise Suppression, Noise Gate and Gain Filter." (n.d.-b). Retrieved March 10, 2019, from https://obsproject.com/forum/resources/obs-studio-how-to-configure-your-microphone-noise-suppression-noise-gate-and-gain-filter.423/
55. Podigee. (n.d.). "How Podcasters Can Use Twitch to Grow Their Audience." Retrieved March 10, 2019, from https://www.podigee.com/en/blog/how-podcasters-can-use-twitch-to-grow-their-audience/
56. politizane. (n.d.). *Wealth Inequality in America*. Retrieved from https://www.youtube.com/watch?v=QPKKQnijnsM
57. "PTZOptics Camera Controller for OBS." (n.d.). Retrieved March 10, 2019, from https://obsproject.com/forum/threads/ptzoptics-camera-controller-for-obs.84504/
58. reznoire. (n.d.). *Double Fine Adventure Kickstarter Promotional*. Retrieved from https://www.youtube.com/watch?v=uYZ_RnPMlQw
59. Richards, P. (n.d.). "How to Engage with Your Audience with Live Video [Customer Story]." Retrieved March 10, 2019, from https://blog.hubspot.com/customers/how-to-engage-your-audience-with-live-video
60. Robson, D. (n.d.). "Our fiction addiction: Why humans need stories." Retrieved March 8, 2019, from http://www.bbc.com/culture/story/20180503-our-fiction-addiction-why-humans-need-stories
61. "Seamless way to setup a second streaming PC." (n.d.). Retrieved March 10, 2019, from https://obsproject.com/forum/resources/seamless-way-to-setup-a-second-streaming-pc.670/
62. "Setting up Live YouTube chat that updates with new streams | Open Broadcaster Software." (n.d.). Retrieved March 10, 2019, from https://obsproject.com/forum/resources/setting-up-live-youtube-chat-that-updates-with-new-streams.591/
63. "Shifts for 2020: Multisensory Multipliers." (n.d.). Retrieved March 8, 2019, from https://web.facebook.com/business/news/insights/shifts-for-2020-multisensory-multipliers
64. "Storytelling to make an impact - YouTube." (n.d.). Retrieved March 10, 2019, from https://creatoracademy.youtube.com/page/lesson/storytelling
65. "Stream Effects | Open Broadcaster Software." (n.d.). Retrieved March 10, 2019, from https://obsproject.com/forum/threads/stream-effects.76619/
66. "Stream/Recording Start/Stop Beep (SRBeep)." (n.d.). Retrieved March 10, 2019, from https://obsproject.com/forum/resources/stream-recording-start-stop-beep-srbeep.392/

67. "The future of live streaming video: today's top trends and the live video of tomorrow." (2018, August 9). Retrieved March 8, 2019, from https://www.epiphan.com/blog/future-of-live-streaming-video/
68. "The Psychology of Storytelling." (2012a, October 16). Retrieved March 8, 2019, from https://www.sparringmind.com/story-psychology/
69. "The Psychology of Storytelling." (2012b, October 16). Retrieved March 8, 2019, from https://www.sparringmind.com/story-psychology/
70. "What consistency can do for your brand, with Vincenzo Landino." (2016, November 10). Retrieved March 10, 2019, from https://livestreamingpros.com/vincenzolandino/
71. "Which Streaming Video Software Should I Use? OBS vs. vMix vs. Wirecast." (2018, October 25). Retrieved March 10, 2019, from https://www.dacast.com/blog/streaming-video-software/
72. "Why You Need to Start Thinking About Mobile Learning." (n.d.). Retrieved March 10, 2019, from https://www.go1.com/blog/post-need-start-thinking-mobile-learning
73. "Why You Should Care About Live Streaming in 2018." (2017, September 21). Retrieved March 8, 2019, from https://neilpatel.com/blog/live-streaming-importance-2018/
74. Wolf, D. (n.d.). "Educational Benefits and Opportunities of Live Streaming | Emerging Education Technologies." Retrieved March 10, 2019, from https://www.emergingedtech.com/2017/04/educational-benefits-and-opportunities-of-live-streaming/
75. "Yahoo:_The Live Video Opportunity"_2016.pdf. (n.d.). Retrieved from https://admarketing.yahoo.net/rs/118-OEW-181/images/Yahoo_The%20Live%20Video%20Opportunity_2016.pdf
76. Bowman, Matt. Forbes, 2017. Retrieved from https://www.forbes.com/sites/forbesagencycouncil/2017/02/03/video-marketing-the-future-of-content-marketing/#63c6aced6b53
77. Cole, Nicholas. July, 2018. If You're Doing Content Marketing: Do You Want Vanity Or Do You Want Exposure? Retrieved from: **https://medium.com/@nicolascole77/if-youre-doing-content-marketing-do-you-want-vanity-or-do-you-want-exposure-35240b3b6f1f**

NDI is a registered trademark owned by NewTek, Inc.

Printed in Great Britain
by Amazon